Contents

Unit AS 1:
Forces, Energy and Electricity

1.1 Physical Quantities

Students should be able to:

1.1.1 describe all physical quantities as consisting of a numerical magnitude and unit;

1.1.2 state the base units of mass, length, time, current, temperature, and amount of substance and be able to express other quantities in terms of these units;

1.1.3 recall and use the prefixes T, G, M, k, c, m, μ, n, p and f, and present these in standard form;

Physical Quantities

To describe a physical quantity we first define a characteristic unit. To state a measurement of a physical quantity, such as force, we need to state two things:

1. A **magnitude** (a numerical value) and

2. A **unit**.

International System of Units (SI units)

The SI system of units defines seven base quantities from which all other units are derived. The table below shows the **six base quantities and the units in which they are measured**.

Quantity	Unit	Symbol
mass	kilogram	kg
time	second	s
length	metre	m
electric current	ampere	A
temperature	kelvin	K
amount of substance	mole	mol

Multiples and submultiples of these base units are commonly used:

Prefix	Factor	Symbol
femto	10^{-15}	f
pico	10^{-12}	p
nano	10^{-9}	n
micro	10^{-6}	μ
milli	10^{-3}	m
centi	10^{-2}	c
kilo	10^{3}	k
Mega	10^{6}	M
Giga	10^{9}	G
Tera	10^{12}	T

Derived Units

Many SI units are **derived**. They are defined in terms of two or more base units. For example velocity, in metres per second, which is written as $m\ s^{-1}$. You must be able to write a physical quantity in terms of its base units. Note that a linear format should be used to express (base) units, ie there should be no solidus (slash) in the units. For example, metres per second should be written as $m\ s^{-1}$ and not as m/s. (CCEA GCE Physics Summer Series 2016 Chief Examiner's Report).

Worked Example

The derived unit for energy is the joule. What are the base units of energy?

To calculate the base units for energy we can use any valid formula for energy such as that below for kinetic energy, E_k:

E_k = ½mv² (the ½ being a number has no units)

unit for energy = unit for mass × unit for velocity × unit for velocity

= kg × m s⁻¹ × m s⁻¹

= kg m² s⁻²   which gives the joule **in terms of base units only**.

Significant Figures

The number of **significant figures** (sf) in a number is found by counting all the digits from the first **non-zero** digit on the left. A zero between two non-zero digits **is** significant.

For example 12.35 has four significant figures, because we start counting from the 1 which is the first non-zero digit on the left. The number 0.0516 has three significant figures, counting from the 5 which is the first non-zero digit on the left. The leading zeroes are essential to give the magnitude of the number. The value of π to six significant figures is 3.14159.

When you have to perform calculations on a set of measurements then the result should be given to the same number of **significant figures** as the initial values. For example, $3.25^2 = 10.5625$, but this should be quoted as 10.6.

Rounding

Rounding involves reducing the number of significant digits in a number. The result of rounding is a number having fewer non-zero digits, yet be similar in magnitude. The result is less precise but easier to use. For example, 9.51356 rounded to two significant figures is 9.5.

The procedure for rounding is:
• Decide how many significant figures you want. In the example of 9.51356 given above, this is two.
• Decide which is the last digit to keep, in this case the 5.
• Increase it by 1 if the next digit is 5 or more (this is called rounding up).
• Leave it the same if the next digit is 4 or less (this is called rounding down).

Variations in the final answer based on the rounding of intermediate values will **not** cost a candidate marks (CCEA GCE Physics Summer Series 2016 Chief Examiner's Report).

Exercise 1

1. (a) The SI unit of force is the newton. Express the newton in SI base units.

   (b) The SI unit of pressure is the pascal. Pressure is defined as Force ÷ Area. Express the pascal in SI base units.

   (c) Momentum is defined as mass × velocity. What are SI base units of momentum?

2. The magnitude of a physical quantity is quoted in the following way 5.5 kJ ms⁻¹.

   (a) Explain the meaning of each prefix.

   (b) What physical quantity is being quoted in this way?

   (c) What are the SI base units of this physical quantity?

3. The energy, E, of a photon of wavelength, λ, is given by the equation E = hc/λ,

   where c is the speed of light. Find the SI base units in which h is measured.

4. All physical quantities consist of a magnitude and a unit. Express each of the physical quantities below in the unit indicated.

   (a) 23.1 cm in m

   (b) 25 km hr⁻¹ in m s⁻¹

   (c) 3.5 MJ in kJ

1.2 Scalars and Vectors

Students should be able to:

1.2.1 distinguish between and give examples of scalar and vector quantity;

1.2.2 resolve a vector into two perpendicular components;

1.2.3 calculate the resultant of two coplanar vectors by calculation or scale drawing, with calculations limited to two perpendicular vectors;

1.2.4 solve problems that include two or three coplanar forces acting at a point, in the context of equilibrium;

Distinguishing Scalars and Vectors

A **vector** is a physical quantity that needs magnitude, a **unit** and a **direction**.

A **scalar** is a physical quantity that requires only magnitude and a unit.

Here is a table of some of the more common vectors and scalars:

Vector	Scalar
Displacement	Distance
Velocity	Speed
Acceleration	Rate of change of speed
Force	Mass
Momentum	Electric charge
	Kinetic energy
	Temperature
	Area
	Volume
	Time
	Electric Current*

* In addition to having magnitude, direction and a unit, vectors are combined vectorially, for example as the parallelogram of forces. However, electric currents are combined algebraically and so are considered scalar quantities. For example, consider two currents entering a junction: the current leaving is the algebraic sum of the two.

Combining Coplanar Vectors

When we add vectors we have to take into account their direction as well as their magnitude. When we add two or more vectors, the final vector is called the **resultant**. For two forces of 15 N and 10 N acting in the **same direction**, the resultant is 25 N. For two forces of 15 N and 10 N acting in opposite directions, the resultant is 5 N in the direction of the larger force.

Adding and Subtracting Vectors

If the vectors are not in a straight line then we use the **nose to tail** method to find the resultant. In the diagram below the resultant of the two vectors, A and B, is C. **C = A + B**

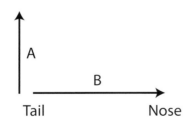

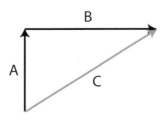

The resultant of subtracting the vector B from A is another vector D. The vector **−B** is a vector of the **same magnitude as B** but in the **opposite direction**. Effectively we add the negative vector so **D = A + (−B)**

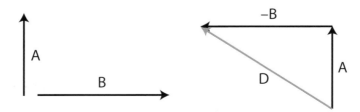

By drawing the vectors to scale and adding them in pairs we can find the resultant of any number of vectors, provided we know their magnitude and direction. This method can be used even if the vectors are not at right angles.

Worked Example

Look at the diagram. Linda moved 3.0 m to the east (AB)
and then 4.0 m to the north (BC). Find the magnitude and
direction of the resultant.

$AC^2 = AB^2 + BC^2 = 3^2 + 4^2 = 25$

$AC = \sqrt{25} = 5.0$

Although she has moved a total distance of 7.0 m,
her displacement is **5.0 m** (AC) from the start. Since
displacement is a vector, a magnitude and a direction are
both needed.

$\tan \theta = $ opposite $\div$ adjacent $= 4.0 \div 3.0 = 1.33$

giving **$\theta = 53°$** (to 2 sf, same as the initial values).

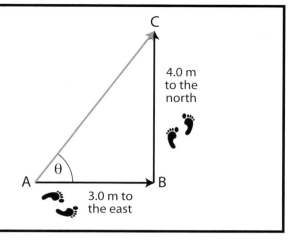

Components of a Vector

It is often useful to split or **resolve** a vector into two parts or components. The diagram shows a vector F that has been resolved into two components that are at right angles to each other.

$\sin \phi = $ opp $\div$ hyp $= Fy \div F$ so, **$Fy = F \sin \phi$**

$\cos \phi = $ adj $\div$ hyp $= Fx \div F$ so, **$Fx = F \cos \phi$**

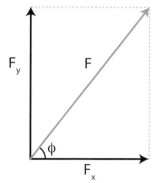

The Inclined Plane

Consider a mass, m, on a plane inclined at angle θ to the horizontal. Its weight, mg, acts vertically downwards as in Diagram A. The component of the weight parallel to the plane is mg sinθ as in Diagram B. The normal reaction to the plane is equal to the component of the weight perpendicular to the plane ($N = mg \cos\theta$) as in Diagram C.

Diagram A

Diagram B

Diagram C

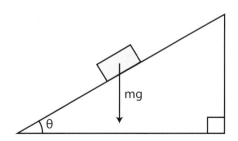
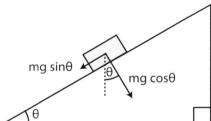
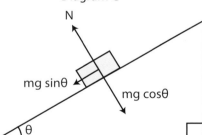

Worked Example

A man, pushing a wheelbarrow and load of total mass
22 kg, approaches a slope inclined at 5.0° to the horizontal,
as shown in the diagram. Calculate the total force the man
must exert on the wheelbarrow and its contents to move it
up the slope at a constant speed of 1.5 m s⁻¹. The frictional
force is constant at 12 N.

(CCEA January 2009, amended)

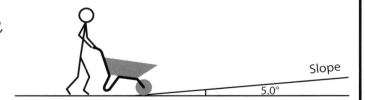

The constant speed means there is no resultant force. Both friction and the component of the weight parallel to the plane act down the slope. The man must therefore exert a force of equal size up the slope.

Force required = Friction + Component of weight parallel to plane

= $12 + 22 \times 9.81 \times \sin 5° = 31$ N (to 2 sf, same as the initial values)

Equilibrium of Forces

If the resultant force on an object is zero then it is in **translational equilibrium.** Consider the worked example below.

Worked Example

Two tugs are used to rescue a small ship which has lost engine power and is close to some rocks. The tugs just manage to hold the ship stationary against a current producing a force of 250 kN on the ship. Tug A develops a force of 200 kN in the direction shown on the diagram. Find the magnitude and direction θ of the force F developed by tug B if the three forces acting on the ship are in equilibrium. (CCEA legacy June 2009, amended)

Since the forces are in equilibrium the vertical components of the forces balance and the horizontal components also balance.

Vertical: **F sin θ** = 200 sin 35   = 115 kN

Horizontal: F cos θ + 200 cos 35   = 250 kN

so   **F cos θ**   = 86 kN

Remembering that sin ÷ cos = tan, we can divide the

equations in bold type to give:

$$\tan \theta = 115 \div 86$$

so $\theta = \tan^{-1}(115 \div 86) = 53°$

Substituting for θ gives:

$$F = 86 \div \cos 53° = 143 \text{ kN}$$

So, F = 143 kN at 53° to the horizontal.

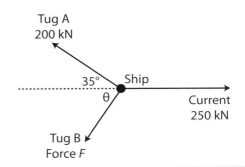

Exercise 2

1. The following is a list of physical quantities:
   Work, Distance, Power, Displacement, Speed, Velocity.

   (a) Underline those that are vectors.

   (b) State the difference between a vector and a scalar.

   (CCEA January 2010 amended)

2. The diagram shows a force of 12 N acting on a brick resting on a horizontal surface. Calculate the resultant horizontal force acting on the brick.

   (CCEA January 2011 amended)

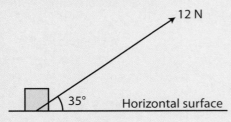

3. The diagram shows two vectors P and Q. Sketch the constructions necessary to obtain the vectors A and B, where A = 2P + 3Q and B = P − 2Q.

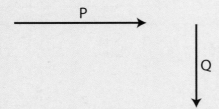

4. The diagram shows two children playing by pulling on a rope connected through a smooth hook on a beam. Child A pulls with a force of 210 N at an angle of 55° from the horizontal.

   (a) Calculate the vertical component of the force with which child A is pulling.

When child B pulls on the rope with a force of 128 N at 20° above the horizontal, the rope does not move.

   (b) What condition must be met for this to happen?

   (c) Confirm, by calculation, that the forces given satisfy this condition.

   (CCEA January 2010 amended)

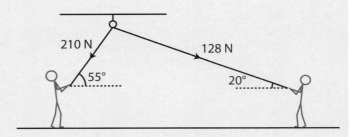

5. Two forces act at right angles to each other as shown. Calculate the resultant force and its direction relative to the vertical.

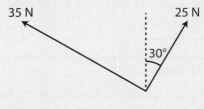

1.3 Principle of Moments

Students should be able to:

1.3.1 define the moment of a force about a point;

1.3.2 use the concept of centre of gravity; and

1.3.3 recall and use the principle of moments;

Moment of a Force

The moment of a force about a point is defined as the product of the force and the perpendicular distance from the point to the line-of-action of the force.

Moment = Force × Perpendicular distance from the point to force

The force is measured in N and the distance in m. Moments are measured in newton–metres, written as N m. The direction of a moment can be clockwise or anti-clockwise. The moment in the diagram below is **clockwise**.

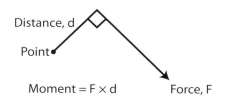

Worked Example

A mechanic tries to remove a rusted nut from fixed bolt using a spanner of length 0.18 m. When he applies his maximum force of 300 N, the nut does not turn.

By placing a steel tube over the handle of the spanner, the length is increased to 0.27 m. When he applies the same maximum force of 300 N at the end of the steel tube, the nut is just loosened.

Calculate the minimum force that would have been necessary to loosen the nut if the length of the spanner had remained 0.18 m.

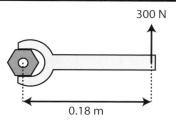

The moment required to loosen the nut is
$300 \times 0.27 = 81$ N m

If the force required to loosen the nut, when the length is 0.18 m, is F.
$F \times 0.18 = 81$
$F = 81 \div 0.18 = 450$ N

Couples

A single force acting on an object will make it move off in the direction of the force. A **couple** is two forces that act in opposite directions, not along the same line, and which cause rotation. A couple produces an **unbalanced moment**.

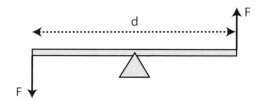

The moment of each force about the pivot is $F \times \frac{1}{2} d$

The sum of these two moments is therefore $F \times \frac{1}{2} d + F \times \frac{1}{2} d = F \times d$

Moment of a couple = One force × Perpendicular separation of the forces

Centre of Gravity and Centre of Mass

The **centre of gravity** of an object is the point at which we can take its **weight** to act.

The **centre of mass** of an object is the point at which we take its **mass** to be concentrated.

A resultant force acting through the centre of mass would cause the object to move in a straight line but without causing it to rotate. For most everyday situations the centre of gravity coincides with the centre of mass.

Principle of Moments

When an object is in rotational equilibrium, the sum of the clockwise moments about any point is equal to the sum of the anticlockwise moments about the same point.

Worked Example

A non-uniform rod of mass 5.50 kg and length 2.00 m is pivoted at a point P at one end of the rod. The rod is held horizontally by a tension of 50.0N acting vertically in a light string fixed to the other end of the rod, as shown. Calculate
(i) the distance of centre of gravity from the point P
(ii) the size and direction of the force acting through point P.

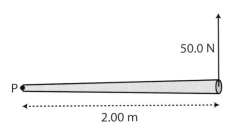

(i) Let the distance from the centre of gravity to the pivot be d.

Then, taking moments about point P,
ACWM = CWM, so $50 \times 2 = (5.50 \times 9.81) \times d$,
so $d = 100 \div 54.0 = 1.85$ m (to 3 sf, same as initial values)

(ii) The weight of the rod is approximately 53.96 N and it acts vertically downwards at the centre of gravity. The total upward forces must balance the total downward forces, so the force through P is **upwards** and of size 3.96 N

Reactions

Many AS questions require the student to **determine the reactions** at points of support. Study carefully the following worked example from an AS paper.

Worked Example

A boy uses a uniform plank of wood of mass 30 kg and length 4.0 m to cross a river. He places one end of the plank on the river bank and rests the plank on a rock in the river as shown in the diagram.

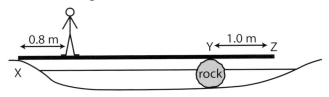

(i) The rock is 1 m from the end of the plank. The boy, who has a mass of 65 kg, stands 0.8 m from the river bank at end X. Calculate the vertical support force, provided by the rock, at Y and the support force provided by the bank, at X.

(ii) Will the boy be able to stand at end Z without the plank rising off the riverbank and the boy falling in the river? Explain your answer.

(CCEA June 2010, amended)

(i) Let the reaction at X be R_x and the reaction at Y be R_y. To find the reaction at Y, take moments about point X. We do this because the support force at X, R_x, has no moment about X and can therefore be ignored **if** we take moments about X.

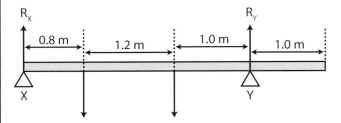

ACWM = CWM **about point X**
$R_Y \times 3 = 30 g \times 2 + 65 g \times 0.8$
$R_Y = 366$ N

ACWM = CWM **about point Y**
$R_X \times 3 = 1 \times 30 g + 2.2 \times 65 g$
$R_X = 566$ N

Observe that the total downward forces,
$95 g = 95 \times 9.81 = 932$ N and the total upward forces
$R_X + R_Y = 566 + 366 = 932$ N.
This provides a convenient check that the reactions have been calculated correctly – it also provides an alternative method to calculate one reaction given the other.

(ii) In the diagram the 30 g (N) force and the 65 g (N) force are both 1.0 m away from Y.

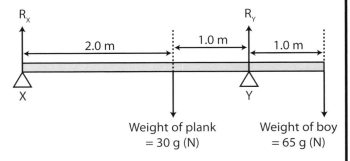

Now, since 65 g > 30 g, there will always be a resultant clockwise moment about point Y. This resultant clockwise moment means that the plank is not in equilibrium and will **tip about Y** by lifting off the bank at X. So the boy is **unable to stand** at point Z without falling into the river.

Exercise 3

1. A stage lighting batten consists of a uniform beam AB, 24 m long, which weighs 600 N. The batten is suspended by two vertical cables C and D.

    The tensions in each cable are equal to 430 N. The batten supports two spotlights S1 and S2 each of weight 70 N and a floodlight F of weight 120 N. The arrangement and distances are shown in the diagram.

    How far is cable C from end A?

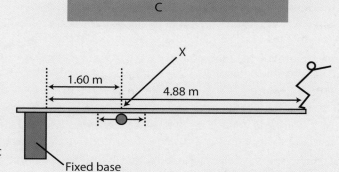

2. A wheel of radius 0.50 m rests on a level road at point C and makes contact with the edge E of a kerb of height 0.20 m, as shown in the diagram. A horizontal force of 240 N, applied through the axle of the wheel at X, is required just to move the wheel over the kerb.

    Find the weight of the wheel.

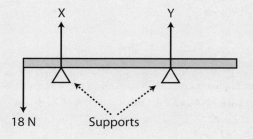

3. A diver stands on the end of an adjustable springboard as shown opposite.

    The diver exerts a moment on the springboard about the pivot at point X.

    (a) On what two factors will the size of the moment the diver exerts depend?

    The total length of the springboard is 4.88 m and the pivot X can be adjusted to move a distance of 0.28 m on either side of its centre position as shown in the diagram.

    (b) (i) Show that the **maximum** moment that a diver of mass 65 kg can exert when she stands on the end of the springboard is 2270 N m.

    (ii) A different diver of mass 75 kg now stands on their own on the end of the springboard. By how much, and in what direction, will the pivot need to be moved from its **central position** for this diver to exert the same moment as the 65 kg diver in (b)(i)?

    (CCEA January 2010)

4. The diagram opposite shows a uniform plank of weight 30 N and length 3 m, resting on two supports. The supports are 0.5 m and 2.0 m from the left hand end of the plank. A weight of 18 N is suspended from the left hand end of the plank.

    (a) Find the reactions X and Y at the two supports.

    (b) By how much should the weight at the left hand end be increased so that the reaction at Y becomes zero?

    (CCEA January 2011)

5. Gymnasts are practising on a uniform wooden beam of weight 124 N and length 180 cm. In order to raise it above the floor, the beam is resting on two metal supports, A and B, each of which is at 20.0 cm from the end of the beam, as shown in the diagram.

    (a) Calculate the maximum weight of gymnast, W, who can stand at the left-hand end of the beam, without the beam beginning to tip up.

    (b) What upward force is provided by the support at A, when this gymnast is standing in this position?

    (CCEA June 2016 amended)

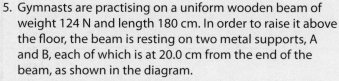

6. (a) A student states the principle of moments as:

    "When an object moves, the sum of the clockwise moments equal the sum of the anticlockwise moments."

    Identify two errors or omissions in the student's statement.

    A person holds a 1.5 kg dumb-bell in his hand and keeps his arm horizontal before lifting the mass upwards. The forearm pivots about the elbow joint and has a weight of 25 N, which acts 19.0 cm from this joint. The force in his bicep F_B acts vertically upwards at a distance of 6.8 cm from the elbow joint.

    The centre of gravity of the dumb-bell is 37.0 cm from the elbow joint. The vertical force at the elbow joint is labelled F_E. The situation is shown in the diagram opposite.

    (b) Using the principle of moments, calculate the magnitude of the force in the bicep F_B.

    (c) State an expression for the vertical force at the elbow joint F_E in terms of the other forces acting when the arm is held horizontal with the dumb-bell in the hand.

    (d) Determine the magnitude of the vertical force acting at the elbow joint F_E.

    (CCEA June 2015)

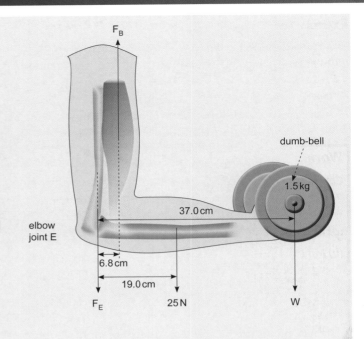

7. An extendable wrench is often used to remove the wheel nuts from a car. The length, L, of the shaft of the wrench can extend from 32 cm to 54 cm as shown opposite.

    (a) Calculate the percentage reduction in the force required to perform the same task with the wheel wrench at its longest compared to when it is at its shortest.

    A 62 kg woman attaches the wrench to a wheel nut and finds it makes an angle of 34° to the horizontal. She finds that by standing on the extreme end of the wrench, which is at its minimum length of 32 cm, she can just loosen the nut attached to the wheel.

    (b) Calculate the moment produced by the woman under these conditions.

    (CCEA June 2014, amended)

1.4 Linear Motion

Students should be able to:

1.4.1 define displacement, velocity, average velocity and acceleration;

1.4.2 recall and use the equations of motion for uniform acceleration;

1.4.3 describe an experiment using light gates and computer software to measure acceleration of free fall, g;

1.4.4 interpret, qualitatively and quantitatively, velocity-time and displacement-time graphs for motion with uniform and non-uniform acceleration;

Definitions

Displacement is the distance moved in a particular direction.

Speed is defined as the distance moved per second.

Velocity is defined as the displacement per second.

Average velocity is defined by the equation: $\text{Average velocity} = \dfrac{\text{Total displacement}}{\text{Total time taken}}$

Acceleration is defined as the rate of change of velocity with time.

Equations of Motion for Uniform Acceleration

On the right are four equations which students need to remember and be able to use.

$$v = u + at$$
$$s = \tfrac{1}{2}(u + v)t$$
$$s = ut + \tfrac{1}{2}at^2$$
$$v^2 = u^2 + 2as$$

where:
u = initial velocity
v = final velocity
a = constant acceleration
t = time taken
s = displacement

Worked Example

Competitor A in a cycle race reaches a point 60.0 m from the finishing line. He then travels with uniform velocity of 18.0 m s⁻¹ in a straight line towards the finish. Another competitor B reaches the same point 60.0 m from the finish 0.100 s after A, travelling with the same velocity (18.0 m s⁻¹) as A. However, B then accelerates uniformly at 0.720 m s⁻² until he reaches the finish.

(a) Calculate the velocity with which competitor B crosses the finishing line.

(b) Make appropriate calculations to determine which competitor wins the race.

(a) $u = 18.0$ m s⁻¹ $a = 0.720$ m s⁻² $v = ?$ $s = 60.0$ m

Consider the motion of B from the 60.0 metre mark.

$v^2 = u^2 + 2as$

$v^2 = 18.0^2 + 2 \times 0.720 \times 60 = 410.4$

$v = \sqrt{410.4} = 20.26 \approx \textbf{20.3 m s}^{-1}$

(b)   For B: $v = u + at$

so $20.26 = 18 + 0.72$

so $t = (20.26 - 18) \div 0.72 = 3.139$ s

But B reached the 60 m mark 0.100 s after A, so B takes a total time of $3.139 + 0.100 = 3.239$ s

A takes $60 \div 18 = 3.333$ s to reach finish, so B wins by 0.094 s

---

Displacement–Time Graphs

Velocity = gradient of the displacement–time graph.

In the first displacement-time graph below, the displacement increases by equal amounts in equal times over the first 10 seconds. This means that the object is moving with **constant velocity**. In the first 10 seconds the velocity is a constant 4 m s⁻¹. It then remains stationary for 4 seconds and finally moves in the **opposite direction** with a constant velocity of 6.67 m s⁻¹ for 6 seconds. The object has finally arrived back at its starting point; the total displacement is zero.

Note carefully that:

1. the gradient of a **distance–time** graph is the **scalar quantity, speed**.

2. the gradient of a **displacement–time** graph is the **vector quantity, velocity**.

The second displacement–time graph below tells us that the velocity of the object is increasing; ie it is accelerating. To find the **actual (instantaneous)** velocity at any time we need to carefully **draw the tangent** to the curve at that time and calculate its gradient. The tangent is a straight line that **touches** the curve but does not cut it.

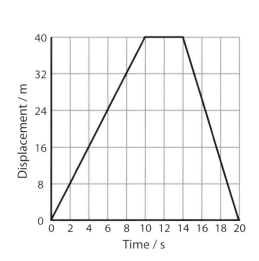

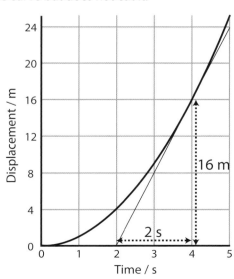

For the graph shown the instantaneous velocity at 4 seconds is the gradient of the tangent to the curve at 4 seconds.
Gradient = 16 ÷ 2 = 8 m s⁻¹.

Note that for an object undergoing uniform acceleration **from rest,** at any given time the instantaneous velocity at that time is always **twice** the average velocity.

Velocity–Time Graphs

Acceleration

This graph shows the motion of an object that is moving in a straight line and always in the same direction. It starts at rest, accelerates from 0 to 10 seconds, travels at constant velocity for 10 seconds, and then decelerates to a stop after a total time of 25 seconds.

The gradient of the line gives us the acceleration or deceleration.

Between 0 and 10 s the velocity change = 12 m s⁻¹.
Gradient = $12 \div 10 = 1.2$ m s⁻²

Between 20 and 25 s the velocity change = –12 m s⁻¹.
Gradient = $-12 \div 5 = -2.4$ m s⁻²

This **negative acceleration could be described as a deceleration** of 2.4 m s⁻².

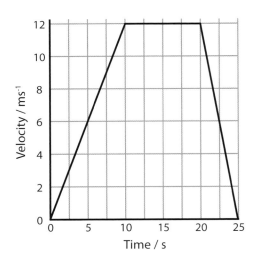

Displacement

From 0 to 10 s:
displacement = average velocity
  × time taken
  = (½ × 12) × 10 = 60 m
  = area of triangle

From 10 s to 20 s:
displacement = constant velocity
  × time taken
  = 12 × 10 = 120 m
  = area of rectangle

From 20 s to 25 s:
displacement = average velocity
  × time taken
  = (½ × 12) × 5 = 30 m
  = area of triangle

So the total distance travelled in 35 s is 60 + 120 + 30 = **210 m**, which is the area of the trapezium. We conclude that **the displacement is the area enclosed between the velocity-time graph and the time axis.**

Note that if the velocity changes from positive to negative it indicates a change in the direction of motion and that has to be taken into account when measuring the displacement. A question in the next exercise provides an example of this.

Variable acceleration

If the object is experiencing a non–uniform acceleration the velocity–time graph is a curve, as shown. As before:
1. the area between the graph and the time axis will give the displacement (first diagram) and
2. the gradient of the curve (Δv/Δt) will give the acceleration at that moment in time (second diagram).

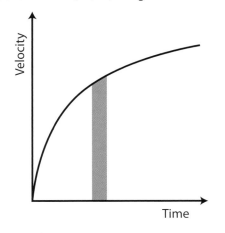

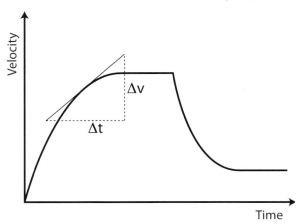

Vertical Motion under Gravity

An object dropped will accelerate due to the force of gravity. We can describe an object as being in **free fall** if the **only** force acting on it is gravity. All objects in free fall accelerate downwards at the same rate independent of the mass of the object. The acceleration due to gravity, g, is approximately 9.81 m s⁻², but its value changes from one point to another over the Earth's surface.

Convention

This book adopts the convention that **upwards** is **positive** when solving problems relating to motion under gravity. This means that an object moving vertically upwards has a positive velocity, while one moving vertically downwards will have a negative velocity. Objects above the surface have a positive displacement, while objects below the surface have a negative displacement.

Worked Example

A ball is dropped from a height of 10.0 m onto a hard surface and bounces to a height of 8.50 m. Calculate the time for the ball to hit the hard surface, the speed with which the ball hits it and the ball's speed immediately after the bounce.

Time to fall, from rest, from 10.0 m:
The ball moves 10.0 m towards the ground, so the displacement is -10.0 m.
Since the acceleration is also downwards, $g = -9.81$ m s^{-2}
$$S = ut + \tfrac{1}{2} at^2$$
$-10.0 = 0 + \tfrac{1}{2} \times (-9.81) \times t^2$ giving $t = \textbf{1.43 s}$

Velocity after falling 10.0 m:
$$v = u + at$$
$v = 0 + (-9.81) \times 1.43$ giving $v = \textbf{−14.0 m s}^{-1}$
where the minus sign shows the ball is moving **towards the ground**.

Initial velocity needed to reach a height of 8.50 m:
The ball is moving upwards, so the displacement and velocity are both positive. But the acceleration due to gravity, g, is towards the ground and is therefore negative.
$$v^2 = u^2 + 2aS$$
$0 = u^2 + 2 \times (-9.81) \times 8.50$ giving $u = \textbf{12.9 m s}^{-1}$

Measuring the acceleration of free fall g.

Method 1 Based on the equation of motion v = u + at

A transparent plastic strip with two opaque strips is dropped vertically through the light gate as shown.

The width of each opaque strip, W, is measured. Each strip should have the same width.

The time for first strip to pass through the beam, t_1, is measured and this allows the initial velocity, u, to be calculated $u = W \div t_1$.

The time for the second strip to pass through, t_2, is also measured and this allows the final velocity, v, to be calculated $v = W \div t_2$.

The time interval, T, between the first strip interrupting the beam and the second strip interrupting is also measured by this method.

Substitution of these values into the equation of motion $v = u + at$ allows g to be calculated: $g = (v - u)\, T$

It is good practice to repeat the experiment and obtain an average for g.

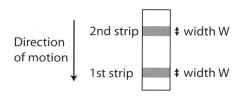

Method 2 Based on the equation of motion v² = u² + 2as

A transparent plastic strip with one opaque strips is dropped vertically through the light gate as shown. The width of the strip, W, is measured.

The strip is dropped from rest a measured distance, s, above the light gate as shown. The initial velocity u is therefore 0.

The time, t, for which light beam is interrupted is measured.

The final velocity is calculated from $v = W \div t$.

The values of u (0), v (W/t) and s are substituted into the equation of motion $v^2 = u^2 + 2gs$ and g calculated. The experiment should be repeated with the strip dropped from different height and an average value of g calculated.

Alternatively a graph of v^2 (*y*-axis) is plotted against s. The graph is a straight line passing through the origin and the gradient equals 2g.

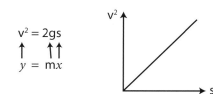

Exercise 4

1. A typical jet aircraft accelerates from rest for 28.0 seconds, leaving the ground with a take-off speed of 270 kilometres per hour.

   (a) Show that a speed of 270 kilometres per hour corresponds to a speed of 75.0 m s^{-1}.

   (b) Assuming that the acceleration of the aircraft is constant, calculate the acceleration of the aircraft as it travels along the runway and the minimum distance the aircraft travels along the runway before it takes off.

   (CCEA January 2008, amended)

2. An athlete runs a hundred-metre race. He accelerates uniformly from rest for the first 40.0 m. He then continues to run the remainder of the race at the velocity attained after the initial period of acceleration. He completes this final part of the race in 4.62 s. Calculate the total time taken by the athlete for the race.
(CCEA January 2007)

3. Below is a velocity–time graph for a car travelling in a straight line along a level road.

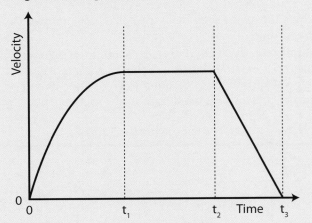

Using the terms **uniform**, **non-uniform**, **zero**, with the words **acceleration** and/or **deceleration**, as appropriate, describe the acceleration or deceleration of the car in the three time intervals indicated.
(CCEA January 2009, amended)

4. A hotel lift, initially at rest, moves vertically in a straight line. From time $t = 0$ it accelerates uniformly until $t = t_1$, when its acceleration suddenly decreases to zero. It continues moving until $t = t_2$ with zero acceleration. It then decelerates uniformly coming to rest at $t = t_3$. The graph below shows how its acceleration depends on time t.

Sketch a graph to show the variation with time t of the velocity v of the lift.
(CCEA January 2008, amended)

5. A ball is thrown vertically upwards with an initial velocity of 39.24 m s^{-1}.

(a) Write down (no calculations required) its speed and its acceleration when it reaches maximum height.

(b) Calculate the maximum height the ball reaches.

(c) How long does it take the ball to reach maximum height?

6. A stone is dropped from rest down a well. Exactly 5.00 seconds after the stone is dropped, a splash is heard.

(a) At what velocity did the stone enter the water?

(b) Calculate the average speed of the stone as it fell.

(c) How far did the stone travel before it hit the water?

7. From the top of a tower 30.0 m high, a marble is thrown vertically upwards with an initial speed of 12.0 m s^{-1}. Calculate:

(a) the maximum height reached above the ground.

(b) the time taken for the stone to reach maximum height.

(c) the time taken for the stone to fall from its maximum height to the ground.

(d) the speed of the stone when it strikes the ground.

8. A helicopter is at a height of 22.0 m and is rising vertically at 4.00 m s^{-1} when it drops a food parcel from a side door.

(a) Write down the velocity and acceleration of the parcel at the instant it leaves the helicopter and then calculate:

(b) the maximum height reached by the parcel before it starts to fall towards the ground.

(c) the velocity of the parcel on impact with the ground.

(d) the time between the parcel leaving the helicopter and it striking the ground.

9. A ball bearing is rolled up a frictionless ramp. It is released with an initial velocity of 2.5 m s^{-1}.

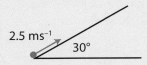

(a) Calculate the deceleration of the ball bearing.

(b) Calculate the distance the ball travels up the ramp before it momentarily stops.

(c) Sketch the velocity-time graph and the displacement-time graph for the motion of the ball-bearing. Your graphs should show appropriate values of velocity, displacement and time.

10. The data below were obtained during an experiment to measure the value of g, the acceleration of free fall. The time taken, t, for a small metal sphere to fall freely from rest through a distance, s, was measured.

s / m	t / s
0.40	0.29
0.60	0.36

Use all the data to calculate a value for g.

(b) What apparatus was required to obtain the data in the table above?

(c) Suggest a possible source of error associated with this experiment, which may lead to calculated values for g which are not equal to the accepted value.
(CCEA June 2016 amended)

11. A new sports car is to be tested on a long, straight, horizontal test track.

(a) The test driver starts the vehicle from rest and accelerates uniformly to a maximum velocity of 200 km h^{-1} in 12.0 seconds. The car continues at this velocity for 20 seconds before the brakes are applied causing constant, rapid deceleration. The car comes to rest in a further 8 seconds.

(i) Show that 200 km h^{-1} is equivalent to 55.6 m s^{-1}.

(ii) sketch a velocity–time graph to represent the motion of the car on the track. Include relevant numerical data on both axes.

(b) Use your graph to find the total distance the car has travelled during this test run.

12. A passenger jet airliner has a landing velocity of 72.0 m s^{-1} as its wheels touch the runway. Its velocity is reduced to its taxiing velocity of 8.50 m s^{-1} in 12.0 seconds as it travels along the runway.

(a) Show that the jet reaches its taxiing velocity in a distance of 483 m.

(b) The airliner must attain a speed of 80.0 m s^{-1} from a standing start to be able to lift off. If the acceleration of the airliner under these conditions is considered constant at 0.96 m s^{-2}, calculate by how much the 2780 m long runway is short.

1.5 Dynamics

Students should be able to:

1.5.1 describe projectile motion;

1.5.2 explain projectile motion as being caused by a uniform velocity in one direction and a uniform acceleration in a perpendicular direction;

1.5.3 apply the equations of motion to projectile motion, excluding air resistance;

Projectiles

A **projectile** is any object that is freely moving in the Earth's gravity. However, what follows relates to motion in which there is both a horizontal and a vertical component of velocity. In other words, Section 1.4 related to motion in a single direction; what follows relates to motion in a plane.

Horizontal projection over a cliff

When the projectile leaves the edge of the cliff it begins to fall vertically. Its downward acceleration is 9.81 m s^{-2}. It is treated as an object dropped vertically from rest so that, at any instant, the velocity of the projectile is the resultant of:
(a) the constant horizontal velocity,
(b) the vertical velocity gained as it falls.

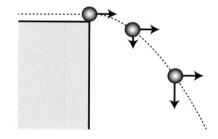

Projectile fired at an angle to the horizontal

Suppose a particle is projected with a velocity u at an angle θ to the horizontal as shown in the diagram below.

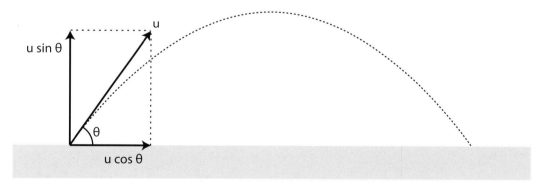

Note carefully that:

1. The **horizontal** component, u cos θ, does not change throughout the motion.

2. The **vertical** component (initially u sin θ), decreases as the projectile moves upwards and increases as it moves downwards.

3. At the maximum height the vertical velocity is momentarily **zero**.

4. The total time in the air is called the **time of flight** and equals twice the time to reach the maximum height.

5. The horizontal distance travelled is called the **range**. Horizontal range = constant horizontal velocity times the time of flight.

6. At any instant the velocity of the projectile is the **resultant** of the constant horizontal velocity and the changing vertical velocity.

Worked Example

A stone is projected into the air from ground level with a velocity of 25.0 m s⁻¹ at an angle of 35.0° to the horizontal.

Calculate:
(a) the time to reach the maximum height
(b) the maximum height reached
(c) the magnitude and direction of the stone's velocity 2.00 s after it was released
(d) the horizontal range.

(a) At the maximum height the vertical component of the projectile's velocity is zero.
The initial vertical velocity = $25.0 \sin 35° = 25.0 \times 0.5736 = 14.34$ m s⁻¹.
Using $v = u + at$ we get $0 = 14.34 + (-9.81) t$
$t = 14.34 \div 9.81 = 1.46$ s = time to reach the maximum height

(b) The maximum height can be found by considering the vertical motion.
$S = ½ (u + v)t = ½ (14.34 + 0) \times 1.46 = 10.5$ m

(c) The horizontal component remains constant throughout the motion. Throughout the motion the horizontal component of the projectile's velocity is $25 \cos 35° = 20.48$ m s⁻¹.
After 2.00 s the vertical component can be calculated using $v = u + at$ where $u = 14.34$, $a = -9.81$ and $t = 2.00$.
This gives $v = -5.28$.
The minus is important, because it tells us that the projectile is now moving down with a velocity of 5.28 m s⁻¹.
Projectile's velocity at $t = 2.00$ s $= \sqrt{20.48^2 + (-5.28)^2} = 21.2$ m s⁻¹
The angle of the velocity to the horizontal is θ where $\tan \theta = 5.28 \div 20.48$ giving $\theta = 14.5°$

(d) Time of flight = 2 × time to reach the maximum height = 2 × 1.46 = 2.92 s
Range = Constant horizontal velocity × Time of flight
Range = 20.48 × 2.92 = 59.8 m

Range

The range, R, of a projectile having an initial velocity, u, at an angle θ to the horizontal is given by:

R = Horizontal velocity × twice the time to reach the maximum height

$R = (u \cos \theta \times u \sin \theta \times 2) \div g$

$R = 2(u^2 \cos \theta \times u \sin \theta) \div g$

For a given initial velocity, the range has a maximum value when the angle of projection is 45°.

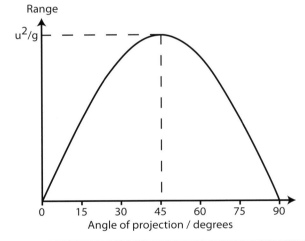

Exercise 5

1. A rugby ball is kicked over the crossbar between the goal-posts from a position 25 m directly in front of the posts, as shown.

   The ball reaches maximum height H above the ground at a position vertically above the crossbar. It takes 1.4 seconds to reach this maximum height. Assume air resistance is negligible.

   (a) Calculate the horizontal and vertical components of velocity at the instant the ball leaves the kicker's foot.

   (b) (i) Use your answers to part (a) to find the magnitude of the initial velocity after the ball is kicked.
   (ii) Find the angle above the horizontal at which the ball is kicked.
   (iii) Find the maximum height H reached by the ball.

   (CCEA June 2009, amended)

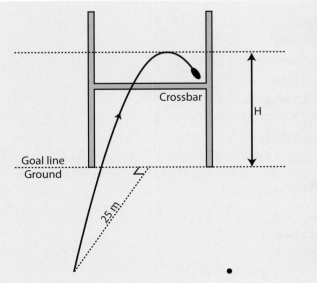

2. When a long jumper completes a jump, his centre of mass is 1.0 m above ground level at take off and 0.4 m above ground level on landing.

   The long jumper has studied some physics and has read that in order to make the horizontal distance of his jump as long as possible he should jump at an angle of 45° to the horizontal. The athlete takes off at a speed of 9.3 m s⁻¹ at an angle of 45° to the horizontal.

Take off point   Landing point

1.0 m   0.4 m

(a) Calculate the initial vertical velocity of the athlete.

(b) Show that the vertical component of the velocity of the athlete on landing has magnitude 7.4 m s⁻¹.

(c) Calculate the time spent in the air by the athlete during the jump.

(CCEA June 2011)

3. A projectile is fired at a target that is 36 m away. The target is on the same horizontal plane that the projectile is fired from. The projectile is projected at 40° to the horizontal. You can assume that the flight of the ball is unaffected by air resistance.

(a) Write **two equations** for the total time taken, T, for the ball to reach the target, one each for the vertical and horizontal components of velocity.

(b) The projectile hits the target. Calculate the initial velocity of the projectile.

4. A projectile is fired at 30° to the horizontal. The graph shows how the vertical component of its velocity changes with time.

(a) Calculate the initial velocity of the projectile.

(b) Calculate the maximum height reached by the projectile.

(c) Calculate the range (maximum horizontal distance travelled) of the projectile.

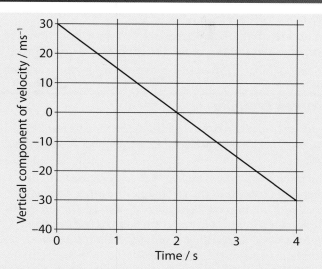

5. An athlete is taking part in a shot-put event. The shot leaves the athlete's hand at a height of 2.0 m above the ground and the velocity of the shot at the instant it leaves the athlete's hand is 13.5 m s⁻¹ at an angle of 40° to the horizontal. The path of the shot is shown in the diagram. Air resistance can be neglected.

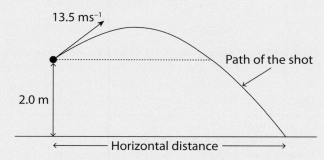

13.5 ms⁻¹

Path of the shot

2.0 m

Horizontal distance

(a) Calculate the time taken for the shot to reach its maximum height.

(b) Calculate the maximum height reached by the shot **above the ground**.

(c) Calculate the horizontal distance travelled by the shot. See the diagram.

(CCEA June 2014)

1.6 Newton's Laws of Motion

Students should be able to:

1.6.1 state Newton's laws of motion;

1.6.2 apply the laws to simple situations;

1.6.3 recall and use the equation $F = ma$, where mass is constant; and

1.6.4 demonstrate an understanding that friction is a force that opposes motion;

Newton's First and Second Laws of Motion

Newton's First Law of Motion
If a body is at rest, it will remain at rest unless a resultant force acts on the object. If the body is moving in a straight line with a constant speed, it will continue to move in this way unless a resultant force acts on it.

Newton's Second Law of Motion

The acceleration of an object is inversely proportional to its mass, directly proportional to the resultant force on it and takes place in the same direction as the unbalanced force. Newton's Second Law can be written: F = ma.

The unit of force, the **newton**, is defined as the force needed to cause a mass of 1 kg to have an acceleration of 1 m s^{-2}.

Worked Example

A person stands on a skateboard at the top of a rough sloping track. The total mass of the rider and skateboard is 73 kg.

The track slopes at an angle of 9.5° to the horizontal. The rider and skateboard start from rest and move down the track with uniform acceleration of 0.46 m s^{-2}. During the motion the force of friction on the board is constant and air resistance is negligible. A simplified diagram of the situation, in which the rider and skateboard have been replaced by a point mass in contact with the track, is shown below.

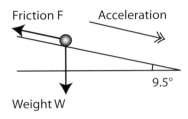

Calculate (a) the magnitude of the resultant force causing the skateboard and rider to accelerate down the slope and (b) the constant frictional force acting on the skateboard as it moves down the slope.

(a) The resultant force is found using
F = ma, where F is the resultant force.
F = 73 × 0.46 = **33.6 N**

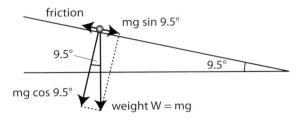

(b) Resultant force = mg sin 9.5° – Friction
33.58 = 73 × 9.81 × 0.165 – Friction
33.58 = 118.20 – Friction
Friction = **85 N** (to 2 sf to match initial values)

Newton's Third Law of Motion

If body A exerts a force on body B, then body B exerts a force of the same size on body A, but in the opposite direction.

The forces of an action–reaction pair always **act on different bodies**.

Two forces that act on the same body are not an action–reaction pair, even though they may be equal in magnitude but opposite in direction.

Consider an apple is in contact with a table. The apple exerts a downward force F_{AT} (AT meaning apple on table): call this the action force. The table exerts an upward force on the apple, F_{TA} (TA meaning table on apple): call this is the reaction force. This is an example of the action–reaction pair of forces to which Newton's Third Law refers.

If we look at the forces acting only on the apple, we have F_{EA} (weight of the apple) and F_{TA} the upward supporting force from the table.

The forces F_{EA} and F_{TA} are equal in magnitude and opposite in direction but they act on the same body, the apple. These do not constitute an action–reaction pair because they act on the same object, in this case the apple.

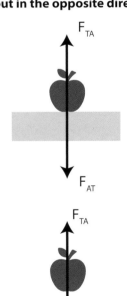

Worked Example

*A toy car of mass 400 g is placed on a slope inclined at 25.0°
to the horizontal. The diagram shows a simplified diagram
of the situation.*

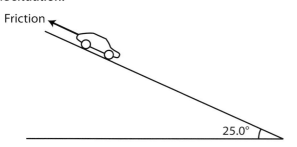

Friction

25.0°

*(a) The car is released and starts to move down the slope.
The force of friction opposing the motion is 1.20 N.
Calculate the initial acceleration of the car down the
slope.*

*(b) The force of friction increases as the car speeds up. At
one stage in the motion the frictional force is increased
by 0.46 N. Describe how this affects the acceleration.
(CCEA January 2008)*

(a) $F_{resultant}$ = mg sin θ – Friction
$= 0.4 \times 9.81 \times \sin 25° - 1.2 = 0.46$ N
acceleration = $F_{resultant} \div$ mass $= 0.46 \div 0.4 = 1.15$ m s^{-2}

(b) As the friction force increases, the resultant force
decreases and the acceleration decreases. When the
friction force reaches 0.46 N, the resultant force is zero
and the car stops accelerating: it moves in a straight line
with constant velocity.

Worked Example

*A man of mass 60.0 kg stands on scales inside a lift. The
scales measure the man's weight, not his mass. What
readings would you expect to see on the scales when the lift
is moving upwards with:*
(a) a constant acceleration of 2.00 m s^{-2}?
(b) a constant speed of 2.00 m s^{-1}?
(c) a constant deceleration of 2.00 m s^{-2}?

The reading on the scales is the reaction force, R, as shown
in each case in the sketches on the right.

(a) When accelerating upwards,
R – mg = ma
R = mg + ma
$= 60 \times 9.81 + 60 \times 2$
$= 709$ N

(b) At constant speed,
R – mg = 0
R = mg $= 60 \times 9.81$
$= 589$ N

(c) When accelerating downwards,
mg – R = ma
R = mg – ma
$= 60 \times 9.81 - 60 \times 2$
$= 469$ N

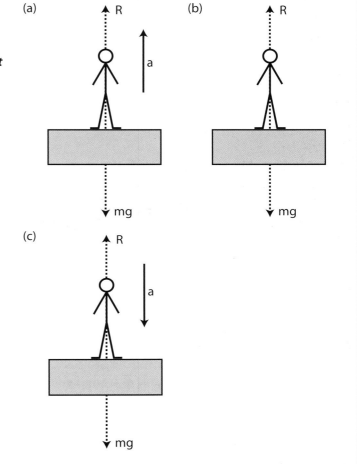

Exercise 6

1. A car of mass of 1800 kg is moving with a speed of
16.7 m s^{-1}. The driver applies a constant braking force
of 1200 N. Calculate the time interval between the
application of the brakes and the car coming to rest.
(CCEA January 2009)

2 (a) A train of mass 2.3×10^6 kg travels at a constant speed
of 20.0 m s^{-1}. There are opposing forces on the train of
0.6 N for every 1 kg of its mass. If the train then

accelerates at 0.20 m s^{-2}, calculate the driving force
required to overcome these opposing forces and
produce this acceleration.

(b) The train now goes onto a section of the track which
has been covered by leaves and is slippery as a result.
Describe and explain what effect this would have on
the motion of the train.
(CCEA January 2010)

3. (a) State Newton's first and third laws of motion.

(b) A student considers a brick resting on the ground as shown below.

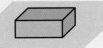

He considered the following four forces which he names forces A, B, C and D.

Force A – The normal contact force exerted by the ground on the brick.

Force B – The weight of the brick.

Force C – The downwards force exerted by the brick on the ground.

Force D – The gravitational attraction of the brick on the Earth.

(i) Referring to the forces above, explain how Newton's first law applies to the brick.

(ii) Referring to the forces above, explain how Newton's third law applies to the brick and the ground.

(CCEA June 2010)

4. (a) State Newton's Second Law of Motion.

(b) A woman of mass 59 kg steps into a stationary lift in order to travel from the tenth floor of an office building to the ground floor. The lift accelerates downwards at 2.5 m s^{-2} until it reaches a steady velocity. It travels at this velocity for a certain time, before decelerating at 2.2 m s^{-2} and coming to rest on the ground floor. The woman feels a reaction force from the floor at all times.

(i) Calculate the size of the reaction force as the woman stands in the lift while it is stationary.

(ii) Calculate the maximum reaction force she will experience.

(c) Describe and explain the circumstance which might lead to the woman experiencing the sensation of 'weightlessness' while standing on the floor of the moving lift.

5. The mass of a sports car and driver is 1480 kg. It is driven along an upward sloping test track inclined at 12° to the horizontal. The driving force F$_D$ from the engine is 8.0 kN as the car accelerates up the slope. The frictional force opposing the motion of the car is 200 N. Calculate the acceleration of the car up the slope.

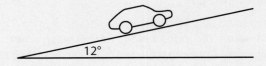

(CCEA June 2015, amended)

1.7 Linear Momentum and Impulse

Students should be able to:

1.7.1 define momentum;

1.7.2 calculate momentum;

1.7.3 apply the principle of the conservation of linear momentum;

1.7.4 perform calculations involving collisions in one dimension;

1.7.5 describe and confirm collisions as elastic or inelastic by calculation;

1.7.6 define impulse as the product F × t;

1.7.7 recall and use the equation Ft = mv – mu;

1.7.8 apply the impulse-momentum relationship to impact situations;

1.7.9 define Newton's second law in terms of momentum;

The momentum of a body is defined as the product of its mass and its velocity. In symbols this is often written:

p = mv   where   p is the momentum in Ns (or kg m s^{-1})
                      m is the mass in kg
                      v is the velocity in m s^{-1}

Momentum is a vector.

Worked Example

(a) Show that the unit Ns is equivalent to the unit kg m s^{-1}.

(b) Calculate the momentum of a car of mass 700 kg travelling due North with a speed of 20.0 m s^{-1}.

(a) From F = ma, we see that the unit of force, the newton, is equivalent to the kg m s^{-2}

So, the Ns = kg m s^{-2} × s = kg m s^{-1}

(b) p = mv = 700 × 20.0 = 14 000 Ns due North

Principle of Conservation of Linear Momentum

The total momentum of a **closed system** remains constant, even during collisions. By a **closed system**, we mean one where no external forces are acting.

Consider the collision of two balls on a snooker table. The collision occurs in an isolated system as long as friction is small enough that its influence upon the momentum of the balls can be neglected. In this case, the only unbalanced forces acting upon the two balls are the contact forces which they apply to one another. These two forces are considered **internal** forces since they result from a source within the system – that source being the contact of the two balls. For such a collision, total system momentum is **conserved**.

However during a collision between two cars on a road, where friction is large, the friction must be considered as an **external** force. This system of colliding objects is not closed and linear momentum is **not conserved**.

This allows us to state the **Principle of Conservation of Momentum** for bodies in collision:

If no external forces are acting, the total momentum of a system of colliding bodies is constant.

Collisions

When we apply this principle to collisions, it can be simply stated as:

Total momentum before collision = Total momentum after collision

---

Worked Example

A toy truck of mass 0.4 kg, moving to the right with a speed of 4.0 m s^{-1}, collides with and sticks to a toy tricycle of mass 1.6 kg moving to the left with a speed of 3.0 m s^{-1}. Calculate:

(a) the momentum of each toy prior to the collision and

(b) the velocity of the combination after the collision.

(a) Taking motion to the right as positive, and motion to the left as negative:

Momentum of truck before collision = mv
mv = 0.4 kg × +4.0 m s^{-1} = +1.6 kg m s^{-1}

Momentum of tricycle before collision = mv
mv = 1.6 kg × −3.0 m s^{-1} = −4.8 kg m s^{-1}

(b) By the Principle of Conservation of Momentum:
Total momentum before collision = Total momentum after collision

{1.6 + (−4.8)} kg m s^{-1} = mass of combination × velocity of combination after collision

$-3.0 = (0.4 + 1.6) \times V_{after}$
$V_{after} = -3.2 \div 2.0 = -1.6$ m s^{-1}

The minus sign shows that the **combined truck and tricycle is moving to the left**, that is, it is moving in the same direction as the tricycle was moving originally.

Worked Example

A bullet of mass 8 g is fired from a pistol of mass 0.8 kg.

If the muzzle velocity of the bullet is 320 m s^{-1}, calculate the recoil velocity of the gun.

Momentum after firing = Momentum before firing

$(0.008 \times 320)_{bullet} + (0.8 \times v_{recoil})_{pistol} = 0$

So: $2.56 + 0.8v_{recoil} = 0$

Hence: $v_{recoil} = -2.56 \div 0.8 = -3.2$ m s^{-1}

The minus sign indicates that the velocities of the bullet and the recoil of the pistol are in opposite directions.

Collision Classification

Collisions may be classified as **elastic** or **inelastic**. **Elastic collisions are those in which kinetic energy is conserved.** These only occur on an atomic scale, such as the collision of two molecules of an ideal gas. **Inelastic collisions are those in which kinetic energy is not conserved.**

Frequently, inelastic collisions involve kinetic energy being converted to thermal energy or sound. An example is the collision of a tennis ball with a racquet. A **completely inelastic** collision is one in which two bodies stick together on impact. Here the loss of kinetic energy is very large, though not complete. An example is a rifle bullet embedding itself in a sandbag.

The **Law of Conservation of Energy** tells us that you cannot just 'lose' energy. Therefore the kinetic energy must have changed into other forms. When the cars collide, a lot of kinetic energy is converted into sound and heat. So the 'missing' kinetic energy has actually been changed into these forms, and work has been done in changing the shape of the cars. But the total energy of the system **does not change**. We can sum up these ideas in a table:

	Momentum	Kinetic energy	Total energy
Inelastic Collisions	is conserved	is NOT conserved	is conserved
Elastic Collisions	is conserved	is conserved	is conserved

Impulse

When a force is applied for a certain amount of time, an impulse is produced. The definition of impulse is the force applied multiplied by the time it was applied. Impulse is calculated by the equation:

Impulse = force × time = Ft   where   F is the force in N
t is the time in s

The units of impulse are Ns or kg m s^{-1}. Impulse is a vector quantity.

When an object experiences an impulse, a change in the momentum of the object results. The change of momentum equals the applied impulse. The relationship between the impulse and the momentum is given by the equation:

Ft = mv – mu   where   m is the mass of the object in kg
u is the initial velocity of the object in m s^{-1}
v is the final velocity of the object in m s^{-1}

The longer the force is in contact with the object, the greater is the momentum change. The term 'follow through' is used in many sports when a ball is struck and is based on the physics of impulse.

Newton's Second Law and Momentum

Newton's second law states that the acceleration a of an object is directly proportional to the resultant force F acting upon the object and inversely proportional to the mass m of the object. We can write this as F = ma.

The acceleration $\mathbf{a} = \dfrac{(v - u)}{t}$ where   u is the initial velocity of the object in m s^{-1}
v is the final velocity of the object in m s^{-1}
t is the time in s

The resultant force $\mathbf{F} = \mathbf{ma} = \dfrac{m(v - u)}{t}$

Observe that the term on the right side of this equation is the rate of change of momentum. So an alternative statement of Newton's second law is that the rate of change of momentum is directly proportional to the resultant force and takes place in the same direction as the force.

Exercise 7

1. A bullet of mass 12 g is fired horizontally from a gun with a velocity of 300 m s^{-1}. It hits, and becomes embedded in, a block of wood of mass 3000 g, which is freely suspended by long strings as shown in the diagram below. Air resistance can be neglected.

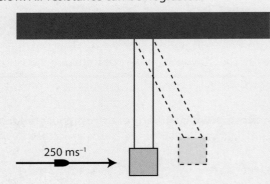

250 ms^{-1}

   (a) Calculate the magnitude of the momentum of the bullet as it leaves the gun.
   (b) Calculate the speed of the wooden block when the bullet strikes it.
   (c) Use your answer to part (b) to calculate the kinetic energy of the wooden block and the embedded bullet immediately after the impact.
   (d) Hence calculate the maximum height above the equilibrium position to which the wooden block, with the embedded bullet, rises after impact.

2. A railway truck T$_1$ of mass 1200 kg is rolling along a track at a velocity of 6.0 m s^{-1} towards a stationary truck T$_2$ as shown in the diagram.

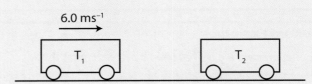

6.0 ms^{-1}

T$_1$ T$_2$

   (a) Calculate the initial momentum of the truck T$_1$.
   (b) On collision, trucks T$_1$ and T$_2$ become joined. They now move with a common velocity of 2.0 m s^{-1}. Find the mass of truck T$_2$.
   (c) Is the collision elastic or inelastic? Your answer should be supported by appropriate calculations.

3. (a) An explosion splits an object, initially at rest, into two pieces of unequal mass. A student observes that the less massive of the two pieces moves with a faster speed than the heavier piece and in the opposite direction. Explain these observations.
   (b) During a ten pin bowling game a player has one pin left to knock down. The player rolls a 7.26 kg bowling ball down the lane and it hits the stationary pin, of mass 1.47 kg, head on at a speed of 8.15 m s^{-1}. After the collision the pin moved forwards at a speed of 13.32 m s^{-1}. Calculate the speed of the ball after the collision.
   (c) State what is meant by an inelastic collision.

(CCEA January 2014)

4. Polonium-210 decays to lead-206 by the emission of an alpha particle as shown in the equation below.

$$^{210}_{84}Po \longrightarrow \, ^{206}_{82}Pb + \alpha$$

(a) Calculate the momentum of the alpha particle if it is emitted with velocity of $+1.60 \times 10^4$ m s^{-1} and a mass of 6.64×10^{-27} kg.

(b) If the polonium nucleus is stationary when the decay occurs, what is the initial velocity of the lead nucleus after the decay? State the direction of motion relative to the α-particle.

(c) State whether this decay is elastic or inelastic and explain your answer with specific reference to this decay.

5. Most military aircraft are fitted with an ejector seat which allows the pilot to escape from the aircraft in case of emergency. One type of ejection system uses an explosion to move the seat, containing the pilot, vertically upwards.

(a) In what direction will the body of the aircraft move as a result of the ejection system being deployed? Explain your answer.

(b) The ejection system is tested in a stationary aircraft on a runway. The mass of the seat is 200 kg and the total mass of the aircraft including the seat is 9100 kg. When the seat is released it leaves the aircraft at a speed of 180 m s^{-1}. In theory, with what initial speed does the body of the aircraft move?

(c) Explain why an explosion such as this can never be considered to be "elastic".

(CCEA January 2012)

6. A rugby player of mass 110 kg moving at a velocity of 2.00 m s^{-1} collides with an opponent moving with twice the momentum in the opposite direction. After the collision the players move together with a common velocity and total kinetic energy of 115 J.

Calculate the magnitude of the common velocity and state the direction the players move after the collision (with respect to the direction the rugby player of mass 110 kg was originally moving).

(CCEA May 2012)

1.8 Work Done, Potential and Kinetic Energy

Students should be able to:

1.8.1 define work done, potential energy and kinetic energy;

1.8.2 show that when work is done energy is transferred from one form to another;

1.8.3 calculate the work done for constant forces, including forces not along the line of motion;

1.8.4 recall and use the equations Δp.e. $= mg\Delta h$ and k.e. $= \frac{1}{2}mv^2$; and

1.8.5 state the principle of conservation of energy and use it to calculate the exchanges between gravitational potential energy and kinetic energy;

1.8.6 use the equation $\frac{1}{2}mv^2 - \frac{1}{2}mu^2 = Fs$ for a constant force;

1.8.7 recall and use: $P = \dfrac{\text{work done}}{\text{time taken}}$, $P = Fv$ and $\text{Efficiency} = \dfrac{\text{useful energy (power) output}}{\text{energy (power) input}}$;

1.8.8 demonstrate an understanding of the importance to society of energy conservation and energy efficiency;

Definitions

- We define the **work done** by a constant force as the product of the force and the distance moved in the direction of the force. Work done = constant force × distance moved in the direction of the force.

  W = F × s

  or **W = Fs cos θ**

  At AS you must be able to apply this new definition when the force and the distance moved are not in the same direction. Study the worked example on page 27.

- A mass has **gravitational potential energy** when it is raised above the ground.

  Δp.e. = mgΔh   where   Δp.e. = change in potential energy in J

  m = mass in kg

  g = acceleration of free fall in m s^{-2}

  Δh = vertical distance in m

  This new form of the equation emphasises the fact that we can only measure the change in gravitational potential energy.

- A moving object possesses **kinetic energy**.

  k.e. = ½ mv²   where   k.e. = kinetic energy in J

  m = mass of the object in kg

  v = speed in m s^{-1}

  The work done by a force accelerating a mass m from a speed u to a speed v is given by: $W = \frac{1}{2}mv^2 - \frac{1}{2}mu^2$

- **Efficiency** is defined as the ratio of useful output work (or power) to total input work (or power).

$$\text{efficiency} = \frac{\text{useful energy (power) output}}{\text{energy (power) input}}$$

Efficiency is a number between 0 and 1 (in accordance with the Law of Conservation of Energy). It has no units.

- **Power** is defined as the rate of doing work. The definition can be expressed by the equation:

$$P = \frac{\text{work done}}{\text{time taken}} \text{ or } P = \frac{\text{energy transferred}}{\text{time taken}}$$

- Since W = Fs and P = W÷t, we have P = Fs÷t. But s÷t = velocity, v. Hence:

P = Fv   where   P = power in watts (W)
F = force being applied in newtons (N)
v = constant speed at which force is moving in m s⁻¹

In recent years examiners have sometimes asked questions relating to **mechanical energy**. Mechanical energy is the sum of the kinetic energy and potential energy. When there is an increase in mechanical energy, work has had to be done by some external agent, such as a car engine. When there is a reduction in mechanical energy, this can be used to do work against frictional forces. In the general case, work is done against friction **and** work is supplied by an external agent.

The general equation is well worth remembering:

Increase in mechanical energy + work done against friction = work supplied from external sources

If the mechanical energy decreases, then we treat it as negative in the equation above.

Energy Conservation and Energy Efficiency

Energy conservation (NOT to be confused with the Principle of Conservation of Energy) is the act of reducing energy consumption, particularly energy used to warm things, light things or move things. Energy conservation has environmental benefits such as reducing the use of fossil fuels which releases large amounts of carbon dioxide into the atmosphere. This contributes to the greenhouse effect, which raises the temperature of the Earth's surface and causes global warming.

Energy efficiency is as any product or process that makes it possible to enjoy the same standard of living while using less energy. This can be achieved in many ways, for example turning down thermostats, washing clothes at low temperatures, taking showers instead of baths, insulating homes, using energy-efficient appliances and turning lights off when not in use.

Worked Example

A cyclist and her machine have a mass of 77.0 kg. She travels a total distance of 800m between two hills. At the top of the higher hill her velocity is 6.20 m s⁻¹. She descends the hill and ascends to the top of the next hill where her velocity is 5.10 m s⁻¹. Throughout the distance travelled the cyclist contends with an average opposing force of 17.0 N.

(a) Air resistance is a possible cause for a force which the cyclist must expend energy to overcome. Suggest two other forces which must also be overcome.

(b) Calculate (i) the total change of mechanical energy of the cyclist between the two hill tops and (ii) the energy contributed by the cyclist for the complete journey between the two hills where the average opposing force of 17.0 N was encountered.

(CCEA January 2007)

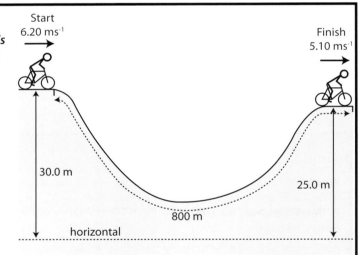

(a) Friction between tyres and the road and the force of gravity (when ascending the hill)

(b) (i) Mechanical energy at start = GPE + KE
    = (77 × 9.81 × 30) + (½ × 77 × 6.2²) = 24 141 J

Mechanical energy at end = GPE + KE

= (77 × 9.81 × 25) + (½ × 77 × 5.1²) = 19 886 J

Total reduction in mechanical energy = 24 141 − 19 886 = 4260 J (to 3 sf)

(ii) Increase in mechanical energy + work done against friction = work supplied from external sources
−4255 + (17 × 800) = energy contributed by cyclist
Energy to overcome resistance forces = F × d = 800 × 17 = 13 600 J
Energy contributed by cyclist (external source) = 13 600 − 4255 = 9350 J (to 3 sf)

Worked Example

An explorer dragging a sledge of mass 70.0 kg across a frozen lake. The explorer attaches the rope to his waist and the force of 200 N is applied at 30° to the horizontal.

(a) How much work is done by the explorer in dragging the sledge 150 metres across the ice at a steady speed of 4.0 m s⁻¹?

(b) Calculate the kinetic energy of the sledge at this speed.

(c) Dragging the sledge over 150 metres generates 4000 J of heat energy and 20 J of sound energy. Calculate the efficiency of the sledge.

(d) Calculate the explorer's useful output power.

(a) The difficulty is that the force, F (200 N), and the displacement, s, are not in the same direction. The easiest solution is to resolve the 200 N force into its vertical and horizontal components as shown below.

Work done = constant force × distance moved in direction of the force = 173.2 N × 150 m = 26 000 J (to 2 sf)

(b) KE = ½ mv² = ½ × 70.0 × 4.0² = 560 J (to 2 sf)

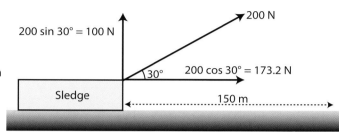

(c) Useful output work = 26 000 J

Total work produced = work against friction + heat + sound = 26 000 + 4000 + 20 = 30 000 J

$$\text{Efficiency} = \frac{\text{useful work output}}{\text{total energy input}} = \frac{26\,000}{30\,000} = 0.867$$

(d) Time taken to drag sledge 150 m at 4 m s⁻¹ is 150 ÷ 4 = 37.5 s

$$\text{Useful output power} = \frac{\text{useful work done}}{\text{time taken}} = \frac{26\,000}{37.5} = 693\,\text{W}$$

Worked Example

The dam at a certain hydroelectric power station is 170 m deep. The electrical power output from the generators at the base of the dam is 2000 MW. Given that 1 kg of water has a volume of 0.001 m³, calculate the minimum rate at which water leaves the dam in m³s⁻¹ when electrical generation takes place at this rate.

In 1 s, the potential energy converted to electrical energy is 2 × 10³ MJ = 2 × 10⁹ J

Change in gravitational Δp.e. = mgΔh
= m × 9.81 × 170 = 1667.7 × m

So mass removed from dam every second is
(2 × 10⁹) ÷ 1667.7 ≈ 1.20 × 10⁶ kg

Since each kilogram of water has a volume of 0.001 m³, the rate of flow is:
(1.2 × 10⁶) × 0.001 = 1200 m³s⁻¹

Exercise 8

1. A car of mass 1200 kg has an output power of 60 kW when travelling at a speed of 30 m s⁻¹ along a flat road. What power output is required if the same car is to travel at the same speed up a hill of gradient 10%? (Such a hill has a slope angle of tan⁻¹(0.1) or 5.7°.)

2. An electric motor has an output power of 2400 W and is used to raise a ship's anchor. If the tension in the cable is 8 kN, at what constant speed is the anchor being raised?

3. (a) Distinguish between kinetic energy and gravitational potential energy.

   (b) A particle possesses energy in two forms only: kinetic energy and gravitational energy. It has a total energy of 3.0 J and is initially at rest. Its potential energy E_p changes causing a corresponding change in its kinetic energy E_k. No external work is done on or by the system. Copy the grid and draw a graph of kinetic energy E_k against potential energy E_p.

   Explain how your graph illustrates the principle of conservation of energy.

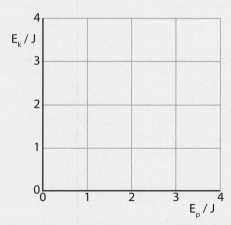

   (c) An AS Physics student plans to enter for the high jump event at the School Sports. She estimates that, if she is to have a chance of winning, she will have to raise her centre of mass by 1.6 m to clear the bar. She will also have to move her centre of mass horizontally at a speed of 0.80 m s⁻¹ at the top of her jump in order to roll over the bar.

27

(i) The student's mass is 75 kg. Estimate the total energy required to raise her centre of mass and roll over the bar.

(ii) The student assumes that this energy can be supplied entirely from the kinetic energy she will have at the end of her run-up. Estimate the minimum speed she will require at the end of the run-up.

(CCEA June 2006)

4. A filament lamp rated 60 W has an efficiency of 0.04 (4%). A modern long-life lamp is rated 12 W and produces the same useful output power as the filament lamp. Calculate

   (a) the useful output power of the filament lamp and

   (b) the efficiency of the long-life lamp.

5. To enable a train to travel at a steady speed of 30 m s⁻¹ along a level track, the engine must supply a pulling force of 50 kN.

   (a) How much work is the engine doing every second?

   (b) If the power is proportional to the cube of the velocity, how much power is needed to drive the train at a speed of 40 m s⁻¹?

6. A cyclist starts from rest at the top of a hill which has a vertical height of 8 m. As she freewheels down the hill, 15% of her energy is dissipated as heat due to friction. The combined mass of the bicycle and cyclist is 90 kg.

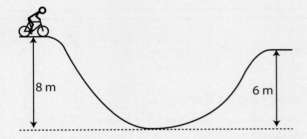

   (a) Calculate the speed with which she reaches the bottom of the hill.

   (b) The cyclist starts to pedal at the bottom of the next hill which is 6 m high. She reaches the top of this hill at a speed of 8.9 m s⁻¹. Assuming there are no energy losses after the cyclist reaches the bottom of the first hill, calculate how much work the cyclist does as she pedals to the top of the second hill.

(CCEA January 2010)

7. The diagram shows a typical tidal barrage. A tidal barrage use the ebb and flow of the tides to produce electrical energy. The moving water drives a turbine and ultimately electrical energy is generated.

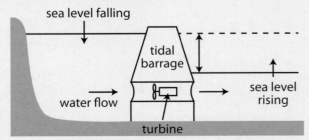

The water falls through a vertical height of 6.0 metres as the tide changes. Calculate the minimum mass of water moving through the turbine every second if the electrical power output from the barrage is 30 MW. Assume no energy losses. Give your answer to 2 significant figures.

(CCEA June 2015)

8. Curling is sport in which players slide a heavy stone across a sheet of ice. During a practice session, a stone launched with a kinetic energy of 19.2 J travels 6.59 m before it stops.

   (a) Determine the magnitude of the average opposing force acting on the stone during its passage across the ice from its launch point.

   (b) The rules of curling allow the ice to be swept. This reduces the size of the average opposing force by 12%. Calculate the kinetic energy with which a stone must be released if it is to stop in 6.59 m if the ice is swept for the last 3.00 m of its journey.

(CCEA June 2014, amended)

9. (a) Define the term **work**.

   A rope exerts a force of 240 N on a box of mass 6 kg to move it at a **steady speed** along the ground. The rope acts at an angle of 28° to the vertical in moving the box a distance of 36 m from Position 1 to Position 2 as shown in the diagram.

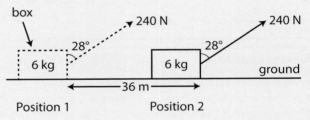

   (b) Calculate the value of the frictional force between the box and the ground.

   (c) Calculate the work done in moving the box from Position 1 to Position 2.

(CCEA June 2013)

10. A skier of mass 78 kg is **stationary** at the top of a 34 m long slope as shown in the diagram.

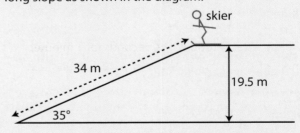

Calculate the velocity attained by the skier during the 34 m descent down the slope if 10% of the skier's energy is used up in overcoming friction.

(CCEA June 2013)

11. A javelin is thrown and fell so that it had a kinetic energy of 245 J as it hit the ground. The tip of the javelin entered the soil and travelled 6.5 cm before coming to a complete stop. Calculate the average resistive force acting on the javelin as it moved through the soil.

(CCEA June 2016, amended)

1.9 Electric Current, Charge, Potential Difference and Electromotice Force

Students should be able to:

1.9.1 recall and use the equation $I = \dfrac{Q}{t}$;

1.9.2 recall and use the equations $V = \dfrac{W}{q}$, $V = \dfrac{P}{I}$;

1.9.3 define the volt;

1.9.4 define electromotive force, E;

1.9.5 distinguish between electromotive force and potential difference;

Conduction in Solids

In metallic conductors the carriers are 'free' electrons. The 'free' electrons in a metal are in rapid, random motion, with **thermal speeds around 1×10^6 m s^{-1}**. Since this motion is random, we do not observe an electric current.

When a battery is applied across the ends of a conductor, the electrons accelerate towards the region of positive potential and to gain kinetic energy. Electron collisions with the vibrating atoms cause the electrons to slow down and give up some of their kinetic energy to the atoms themselves. Externally, we observe this increased internal energy as a temperature rise in the conductor. **Electrical resistance is explained by collisions between the 'free electrons' and the vibrating atoms** of the metal. Following any collision, the electrons accelerate once again and the process continues.

There is a **drift of negative charge** towards the region of positive potential. **It is this drift of electrical charge which constitutes an electric current in a metal.** A typical drift velocity is less than 1 mm s^{-1}.

Current and Charge

By convention, current flows from the region of positive potential to that of lower potential. The current in metals is entirely due to the motion of electrons in the opposite direction to that of the conventional current. The quantity of electric charge flowing past a fixed point is defined in terms of the current.

Thus, for a constant current I flowing for a time Δt we can write:

$\Delta Q = I \times \Delta t$ or $I = \Delta Q \div \Delta t$ where ΔQ = charge flowing past a fixed point, in C (coulombs)
I = constant current, in A (Amperes)
Δt = time taken for charge to flow past fixed point, in s

This tells us that:

1. an electric current is the electric charge passing a fixed point in one second, and

2. a current of 1 A flowing in a circuit is equal to a charge of 1 C passing a fixed point in the circuit every second.

> ### Worked Example
>
> *If the charge on a single electron is -1.6×10^{-19} C, how many electrons flow past a fixed point every minute when a current of 2 A is flowing?*
>
> $\Delta Q = I \times \Delta t = 2 \times 60 = 120$ C
>   (note the substitution of 60 seconds for 1 minute)
>
> Number of electrons = Total Charge $\div$ Charge on a single electron
> $= 120 \div 1.6 \times 10^{-19} = 7.5 \times 10^{20}$ electrons

> ### Worked Example
>
> *What steady current flows when a charge of 300 mC flows past a fixed point in 5 s?*
>
> I $= \Delta Q \div \Delta t = 300 \times 10^{-3} \div 5$
> $= 60 \times 10^{-3}$ A $= 60$ mA

> ### Worked Example
>
> *When a switch in an electrical circuit is closed, a bulb lights instantly, regardless of its distance from the battery. Explain why there is no time delay due to the distance between the lamp and the battery.*
>
> Electron drift begins **instantly** at **all** points in the circuit.

Electromotive Force (e.m.f.)

We picture a battery as a pump which moves electrons from the negative to the positive terminals around a circuit. A battery does **work** on charges, so **energy must be changed within it.** A battery or generator is said to produce an electromotive force (e.m.f.), defined in terms of energy change. The **electromotive force (e.m.f.)** of a battery is defined as **the energy converted into electrical energy when unit charge (1 C) passes through it**:

$$\text{e.m.f.} = \frac{\text{electrical energy converted}}{\text{electric charge moved}} \qquad E = \frac{W}{Q}$$

where E = e.m.f. in V (volts)
W = electrical energy converted in J
Q = charge in C

The unit of e.m.f., like the unit of potential difference (p.d.), is the volt. The volt can be thought of as a joule per coulomb or a watt per ampere.

Remember:

1. The potential difference between two points is the energy converted for every coulomb passing between them.
2. The potential difference between two points is 1 volt if, when 1 coulomb passes between them, 1 joule of energy is converted.

Distinction Between e.m.f. and p.d.

Although e.m.f. and potential difference have the same unit, they deal with different aspects of an electric circuit.

*   **Electromotive force** applies to **a source** supplying electrical energy.
*   **Potential difference** refers to the **conversion** of electrical energy to **other energy forms** by a device in a circuit.

The term 'e.m.f.' is misleading since it measures **energy per unit charge** and not force.

A voltmeter measures potential difference. A voltmeter connected across the terminals of an electrical supply, such as a battery, records the **terminal p.d.** of the battery.

Some people prefer to think of e.m.f. of a battery as **the p.d. across its terminals when no current is drawn from it**. Our definition of the volt allows us to write:

W = QV   where W = work done in J
Q = charge moved in C
V = p.d. in V

All sources of e.m.f. have an **internal resistance** from which the source cannot be separated. When the source provides an electrical current to some external load resistor, a voltage is also developed across this internal resistance. The difference between the e.m.f., E, and the voltage across the external load resistor, V, is equal to the voltage lost in the internal resistor.

Electrical Power

Electrical power is defined as **the rate at which electrical energy is converted into other forms of energy by a circuit or a component, such as a resistor, in a circuit.**

Electrical power, like mechanical power, is measured in watts (W). If we divide both sides of the equation W = QV by time, t, we arrive at:

P = IV   where P = power in W
I = current in A
V = potential difference in V

Worked Example

An electron in a cathode ray tube is accelerated from rest through a potential difference of 150 kV.

(a) Calculate the KE of the electrons when they collide with the screen.

(b) If the current is 32 mA, how many electrons strike the screen per second?

(c) At what rate must heat be dissipated from the screen when it reaches its working temperature?

(a) $W = QV = 1.6 \times 10^{-19} \times 150 \times 10^3 = 2.4 \times 10^{-14}$ J

(b) From the definition of charge, the total charge arriving per second = 32 mC.
Since the charge on each electron is (−) 1.6×10^{-19} C, the number of electrons arriving per second
= $(3.2 \times 10^{-3}) \div (1.6 \times 10^{-19}) = 2 \times 10^{16}$.

(c) $P = IV = 32 \times 10^{-3} \times 150 \times 10^3 = 4800$ W

Exercise 9

1. Terms often used in describing an electrical power source, such as a battery, are **terminal potential difference** and **electromotive force (e.m.f.)**. Write a short explanation of these terms and when it is appropriate to use them.
(CCEA June 2009)

2. A fully charged 12 V battery can deliver 1 A for 20 hours. Calculate for the fully charged battery the total charge stored and the total energy stored.

3. The battery in a mobile phone takes exactly 3 hours to charge when the charging current is 400 mA.
   (a) (i) Calculate the charge which is transferred to the battery in this time.
   (ii) Calculate the number of electrons transferred to the battery in this time.
   (b) Early transmissions of information across the Atlantic Ocean through a submarine cable involved the use of direct current signals. Calculations show that the drift speed of the electrons is such that it would take about 200 years for an electron to cross the Atlantic! Explain why there is no such delay in sending information across the Atlantic in this way.
   (CCEA January 2008)

4. (a) Define electric current.
   (b) 5.0×10^{20} electrons pass normally through a cross-section of a wire in 25 s. Find the current in the wire.
   (c) A number of electrons travel between two electrodes in an evacuated tube. This flow of electrons may be considered to be an electron beam current. The mean speed of the electrons is 8.0×10^6 m s^{-1} and the distance between the two electrodes is 0.45 m. The electron beam current is 1.85 mA.
   (i) Calculate the time taken for an electron to travel between the two electrodes at this speed.
   (ii) Hence calculate the number of electrons in the beam at any instant.
   (CCEA January 2009)

5. (a) Define:
   (i) electrical current,
   (ii) potential difference between two points.
   (b) A steady current of 25 mA flows through a component for 2 minutes. The potential difference across the component is constant at 6.0 V for this time. Calculate:
   (i) the total charge passing through the component in this time.
   (ii) the heat energy dissipated by the component in this time.
   (CCEA June 2016)

6. A constant potential difference of 6.3 V is applied between the ends of a uniform metal wire causing a steady current of 12 mA.
   (a) (i) Deduce the charge passing a point in the wire every second.
   (ii) Determine the amount of energy converted from electrical energy while the charge deduced in (a)(i) passes.
   (b) If the 12 mA current in the wire flows for 90 s, calculate the number of electrons which flow past any point in the circuit.
   (CCEA June 2014)

7. To make a mug of coffee 400 g of water was boiled using an electric kettle. This required 4.11×10^{21} electrons to flow past a point in the electric wires. The kettle is connected to a standard 230 V supply (mains) and is switched on for 126 s.
   (a) Show that the average current flowing during the time the kettle is switched on is 5.22 A.
   (b) Determine the power rating of the kettle.
   (c) Determine the electrical energy transferred during the 126 s of the heating by the kettle.
   (CCEA June 2013)

1.10 Resistance and Resistivity

Students should be able to:

1.10.1 perform experiments to confirm the relationships between current, voltage and resistance in series and parallel circuits;

1.10.2 recall and use the equations for resistors in series and in parallel;

1.10.3 recall and use the equations $R = \dfrac{V}{I}$, and $P = I^2R$;

1.10.4 define resistivity;

1.10.5 recall and use the equation $R = \dfrac{\rho l}{A}$;

1.10.6 perform and describe an experiment to measure resistivity;

1.10.7 demonstrate knowledge and simple understanding of superconductivity;

1.10.8 state Ohm's law;

1.10.9 distinguish between ohmic and non-ohmic behaviour;

1.10.10 perform experiments to determine current-voltage characteristics for metallic conductors, including wire at a constant temperature and the filament of a bulb;

1.10.11 sketch and describe the current-voltage characteristics for a metallic conductor, a diode and a negative temperature coefficient (ntc) thermistor;

1.10.12 sketch and explain the variation with temperature of the resistance of a metallic conductor and a negative temperature coefficient (ntc) thermistor;

1.10.13 perform an experiment to show the variation with temperature of the resistance of a negative temperature coefficient (ntc) thermistor;

Current-Voltage Relationship

The ammeter-voltmeter circuit shown on the right allows us to vary and measure the p.d. V across a bulb and measure the corresponding current I. By replacing the bulb with another component such as length of wire or a thermistor the circuit can be used to obtain voltage and current measurements for that component.

Ohm's Law

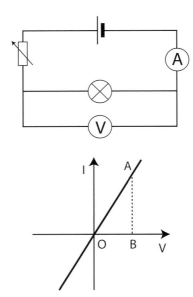

Ohm's law states that the **current through a metallic conductor is directly proportional to the applied p.d., provided the temperature is constant.** These materials are called **ohmic conductors**. We show the relationship as a graph of voltage on the x-axis and current on the y-axis.

Metals and their alloys give I-V graphs which are straight lines through the origin, provided the temperature remains constant. A graph of I against V is called the **characteristic** of the component. Since I is directly proportional to V, it follows that V÷I = a constant. An example of the characteristic for a metallic conductor at constant temperature is shown on the right.

The ratio V/I is called the **resistance**, R of a conductor and is measured in ohms (Ω). Resistance is calculated using the formula:

$$R = \frac{V}{I}$$ where   R = resistance in Ω (ohms)
$\qquad\qquad\quad$ V = potential difference in V
$\qquad\qquad\quad$ I = current in A

We saw in section 1.9 that we were able to write the familiar equation for electrical power, P = IV. We can combine this equation with that for Ohm's law to give:

$$P = IV = I^2R = \frac{V^2}{R}$$

Collectively, these equations are known as **Joule's Law** of electrical heating.

Worked Example

A train of mass 100 000 kg operates from a 25 kV supply and can accelerate to a speed of 20.0 m s⁻¹ in 50.0 seconds along a level stretch of track. Calculate the average current it uses, assuming no energy losses.

Average mechanical power = k.e. ÷ time
= ½ × 100 000 × 20.0² ÷ 50.0 = 400 000 W
I = P÷V = 400 000 ÷ 25 000 = 16 A (to 2 sf)

Worked Example

An electric boiler is rated 2645 W and has a heating element of resistance 20.0 Ω. Calculate the current flowing in the heating element.

P = I²R , so 2645 = I² × 20.0, thus
I=√2645÷20.0=√132.25=11.5 A

Ohmic & Non-Ohmic Behaviour

We have seen that, provided the temperature is constant, the current through a metallic conductor is directly proportional to the applied p.d.,

However, if the temperature of a metal wire is allowed to rise with increasing current, as occurs in the filament of a lamp, then the I-V characteristic curve is as shown on the right. This means that with increasing current (and hence increasing temperature) the resistance of a metal wire increases. However, as the temperature is increasing, the conditions pertaining to Ohm's law are not constant, so we call this **non-ohmic behaviour**.

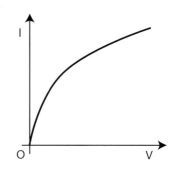

Diodes

A diode is a semiconductor device which has low (ideally zero) resistance to current in one direction and high (ideally infinite) resistance in the other. The most common diode consists of a junction of two types of semiconductor (p type and n type). The circuit symbol for a diode is:

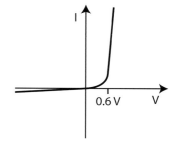

The example I–V characteristic curve on the right shows that this particular diode does not conduct well until the voltage applied exceeds 0.6 V. Beyond this voltage the resistance of the diode is very small. This is an example of **non-ohmic** behaviour.

When a diode is connected to the terminals of a voltage supply with the positive and negative at the ends shown on the circuit symbol, the diode is said to be **forward biased**.

When the diode connections are reversed, so that the end that was previously positive is now connected to the negative terminal of the voltage supply, it is said to be **reverse biased**. If the diode is reversed in the circuit it conducts poorly.

Thermistors

Thermistors are made of semiconductor materials such as silicon or germanium. The I-V characteristic curve for a thermistor is shown on the right. **The resistance of this thermistor decreases as it heats up**. The circuit symbol for a thermistor is:

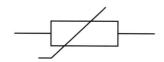

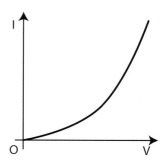

Experiment: How the resistance of an ntc thermistor varies with temperature

The circuit shown on the right can be used to investigate how the resistance of a ntc thermistor varies with temperature. Pour hot water into the beaker so that the thermistor is covered. Record the temperature, voltage and current as the water cools. Alternatively, the experiment can be carried out by starting with a beaker of iced water and then slowly heating it. Calculate the resistance of the thermistor at the various temperatures and plot a graph.

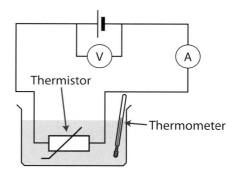

The graph below (left) shows how the resistance of the ntc thermistor decreases with temperature. The **resistance of a ntc thermistor falls exponentially with temperature.** Even at high temperatures a thermistor has a finite resistance, so the graph never touches the horizontal axis.

By comparison, the **resistance of a metal rises linearly with temperature** as shown in the graph below (right). Although there is a linear relationship, **the resistance is not directly proportional to the temperature.** The straight line graph does not pass through the origin, because at 0°C the metal still has resistance.

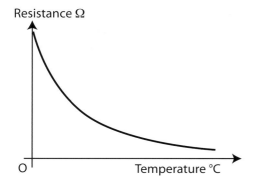

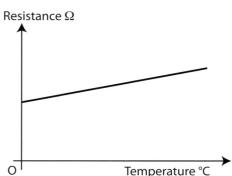

Why do metals and thermistors behave so differently?

When the temperature of a metal increases, the atoms vibrate with greater amplitude and much more violently than before. This causes the free electrons to collide more frequently and with greater force with the atoms in the lattice. These more frequent collisions reduce the average drift speed of the electrons and hence increase the time they take to move between two fixed points in the circuit. This in turn means that, for a given p.d., the current is reduced. Hence the resistance of the metal **increases**.

With thermistors two effects are taking place simultaneously. The increase in temperature would tend to increase the free electron-atom collision frequency and hence increase the resistance, as occurs with metals. However, the electrons in the material of the thermistor are loosely bound to their atoms, and as the temperature increases they become free to move just like the electrons in the metal. This increase in charge carriers more than compensates for the increases in atom-electron collisions and so the resistance decreases.

Current, Voltage and Resistance in Series and Parallel Circuits

Students are expected to perform experiments to measure current and potential difference in series and parallel circuits. Using these measurements they should confirm the relationships used to calculate the total resistance of a number of resistors connected in series and parallel.

Measuring current

An ammeter measures electric current. An ammeter is connected in **series** with the other components in a circuit. The positive terminal of the ammeter is connected to the positive side of the cell. If there are other components in the circuit you should trace the connections from the positive terminal of the cell. The circuits below show how ammeters are connected in series and parallel circuits. In practice you will have access to just one ammeter so the circuit needs to be broken at the position shown in the diagram and the ammeter connected into the gap.

Measuring potential difference

A voltmeter measures potential difference. It is connected in **parallel** with the component across which the potential difference is to be measured. The positive terminal of the voltmeter is connected to the end of the component which is nearest to the positive terminal of the cell, battery or power supply and the negative terminal to the other end of the component. The circuits below show how voltmeters are connected in series and parallel circuits. When building a circuit the voltmeter can be the last item attached.

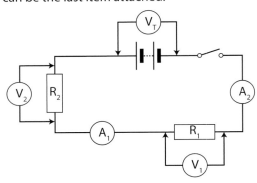

Series circuit   Parallel circuit

In the series circuit ideally you should find that $V_T = V_1 + V_2$. However, in reality the relationship you will find will be close to this but the small differences are due to the fact that voltmeters and ammeters have a resistance of their own, and this alters the total resistance of the circuit and hence the current and potential differences. The ideal voltmeter should have an infinite resistance and the ideal ammeter should have zero resistance.

Resistors in series

The **current** in a series circuit is the same everywhere.

The **supply voltage** is equal to the sum of the voltages across each of the series components:

$V_{battery} = V_1 + V_2 + V_3 + ...$

The **total resistance** is the sum of the resistance of each component:

$R_T = R_1 + R_2 + R_3 + ...$

    $= $ (Battery voltage) $\div$ (Battery current)

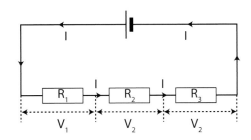

Resistors in parallel

The **potential difference** across each resistor is the same.

The sum of the **currents** through each resistor is equal to the total current taken from the supply.

$$I_{battery} = I_1 + I_2 + I_3 + \ldots$$

The **parallel** resistance formula giving the **total resistance**, R_T is:

$$\frac{1}{R_T} = \frac{1}{R_1} + \frac{1}{R_2} + \frac{1}{R_3} + \ldots$$

When there are **just two resistors** the equation above for R_T reduces to:

$$R_T = \frac{R_1 \times R_2}{R_1 + R_2}$$

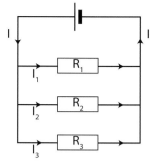

It is important to remember that:

1. the **total resistance** of any **series** arrangement is always **greater than the largest resistance** in that network;
2. the **total resistance** of any **parallel** arrangement is always **less than the smallest resistance** in the parallel network;
3. the **total resistance of a number, N, of equal resistors, R,** arranged in **parallel**, is **R ÷ N**.

Hybrid circuits

Hybrid circuits consist of a mixture of parallel and series elements.

Worked Example

Find the total resistance of the hybrid circuit shown.

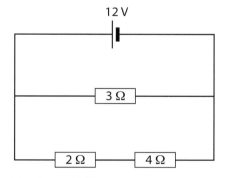

12 V

Applying the series equation first gives the resistance of the 2 Ω and 4 Ω as 6 Ω.

Now applying the equation for parallel networks:

$$\frac{1}{R_T} = \frac{1}{R_1} + \frac{1}{R_2} = \frac{1}{6} + \frac{1}{3} = \frac{1}{2}$$

Hence: $R_T = 2\ \Omega$

We can verify this result by calculating the current in each branch:

The voltage across the 3 Ω resistor = the voltage across the 2 Ω + 4 Ω combination = 12 V

The current in each of the series elements is given by $I = V \div R = 12 \div 6 = 2$ A

The current in the 3 Ω resistor is similarly $12 \div 3 = 4$ A

The current drawn from the battery, $I_b = V_b \div R_T = 12 \div 2 = 6$ A, which is the sum of the currents in the parallel branches of the circuit (2 A + 4 A).

For a current to flow through a conductor there must be a potential difference across its ends. Now consider the resistance networks shown below. The resistance of each of the two networks is 6 Ω and the current taken from each battery is 3 A. In each case the current flowing from X to B is 2 A and the current flowing from X to C is 1 A. In each network, point X is at a potential of 18 V and Y is at a potential of 0 V.

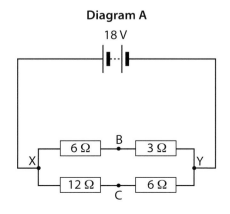

Diagram A

18 V

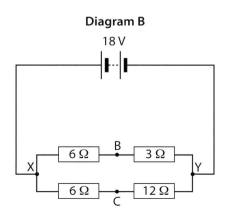

Diagram B

18 V

However:

In Network A

- The p.d. from X to B is 12 V, and the p.d. from X to C is also 12 V.
- Points B and C are both therefore at a potential of 6 V.
- The p.d. between B and C is 0 V.
- **Hence, no current will flow in a short wire of negligible resistance connected between B and C, i.e. because the p.d. is zero.**

In Network B

- The p.d. from X to B is 12 V, and the p.d. from X to C is 6 V.
- Points B and C are at potentials of 6 V and 12 V respectively.
- The p.d. between B and C is 6 V.
- **If a wire of negligible resistance connected between B and C, current will flow from C (higher potential) to B (lower potential).**

Resistivity

Experiment shows that the resistance of a metal conductor at a constant temperature is:
• **directly proportional** its length, L
• **inversely proportional** to its area of cross section, A
This information is represented graphically below.

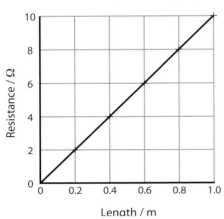

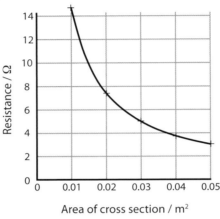

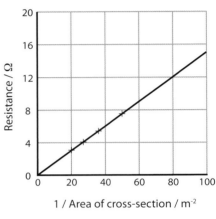

The resistance also depends on the material of the conductor. **Each material has a constant known as its resistivity** and is given the symbol, ρ (pronounced 'rho'). If we combine all these ideas, we have:

$$R = \frac{\rho l}{A}$$

where   R = resistance in Ω
A = area of cross section in m^2
l = length in m
ρ = resistivity in Ωm

The **resistivity of a material** is defined as **numerically equal to the resistance of a sample of the material 1 m long and of cross sectional area 1 m².**
• **Resistivity** is a property of a **material**, such as copper, aluminium, iron etc.
• **Resistance** is a property of a **particular specimen** of a material.
The resistivities of materials vary widely. Metals have resistivity of around 1×10^{-8} Ωm while good insulators have resistivity of around 1×10^{-23} Ωm.

Measuring resistivity experimentally

This is an experiment which is prescribed within the specification. It is important that you do the experiment as part of your training in practical classes and that you can give a detailed description of the procedure for the theory examination. The resistance wire under investigation, which has been previously freed from bends and kinks, is laid along a metre stick and secured in position at each end by means of insulating tape. The electrical circuit is then set up as shown in the diagram above. Connections to the resistance wire are usually made using crocodile clips.

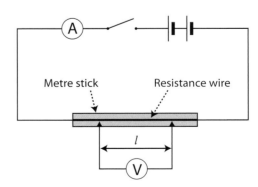

- The experiment involves measuring the voltage across different lengths of wire and the current passing through for lengths l ranging from about 20 cm to about 90 cm.
- From the voltage, V, and current, I, the resistance, R, can be found using R = V ÷ I.
- This is done at least twice per length of resistance wire, and an average taken, to reduce the possibility of random error.

- Plot a graph of R against l. The graph will be a straight line through the origin.
- Now find the gradient of this line to determine the resistance per metre length of wire (R ÷ l).
- Using a micrometer screw gauge we now measure the diameter of the wire at about six points along its length.
- Determine the average diameter <d> and use $A = \pi<d>^2 \div 4$ to find the wire's average cross section area, A.
- The last stage is to calculate the resistivity of the material of the wire, ρ:

$$\rho = \frac{RA}{l} = \frac{R.\pi<d>^2}{4l} = \text{gradient of R-}l\text{ graph} \times \frac{\pi<d>^2}{4}$$

Worked Example

What length of resistance wire must be cut from a reel if the material has resistivity 1.57×10^{-8} Ωm, diameter 0.18 mm and is required to have a resistance as close as possible to 2.50 Ω?

Rearranging $\rho = (R \div l)A$ gives $l = RA \div \rho = R\pi d^2 \div 4\rho$

Hence $l = 2.50 \times \{\pi \times (0.18\times10^{-3})^2\} \div (4 \times 1.57\times10^{-8})$

$l = 4.05$ m

Worked Example

An electric hot plate consists of a 20 m length of manganin wire of resistivity of 4.4×10^{-7} Ωm and cross section area 0.23 mm². Calculate the power of the plate when connected to a 200 V electrical supply.

$P = IV = V^2 \div R = V^2A \div \rho L$
$= (200^2 \times 0.23\times10^{-6}) \div (4.4\times10^{-7} \times 20)$
$= 1045$ W

Superconductivity

The electrical resistance of metals decreases as the temperature falls. For some materials, at a very low temperature, called the **transition temperature,** the resistivity (and hence the resistance) falls to zero. This phenomenon is called **superconductivity**. A graph showing superconductivity in mercury is shown on the right.

We can define a material as a superconductor if it loses all its electrical resistivity to become a perfect conductor when it is below its transition temperature.

Superconductors are used:
• to produce the strong magnetic fields needed for **(MRI) scanners**.
• in **Maglev (magnetic levitation) monorail systems**.

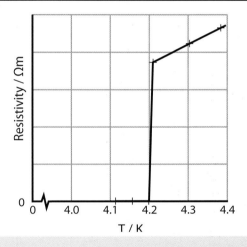

Exercise 10

1. Four identical 6 Ω resistors are connected together in several different arrangements. The diagram below shows one arrangement of the resistors.

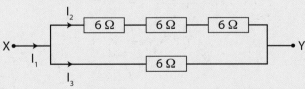

   (a)(i) Calculate the total resistance between terminals X and Y.
   (ii) State the relationship between I_1, I_2 and I_3.
   (iii) The value of current I_1 is 6 A. Determine the current I_3.

   (b) The diagram below shows a different arrangement of the resistors.

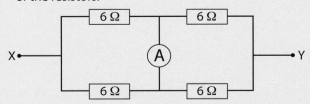

   If the p.d. between X and Y is 10 V, find the current in the ammeter.
   (CCEA June 2010, amended)

2. A battery of e.m.f. 12 V and negligible internal resistance is connected to a resistor network, as shown in the circuit diagram.

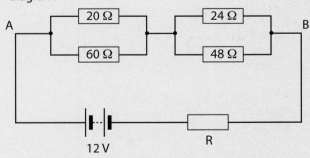

   (a) Show clearly that the resistance of the single equivalent resistor that could replace the four resistors between the points A and B is 31 Ω.

   (b) The current delivered by the battery is 300 mA. Calculate the total circuit resistance.

   (c) Hence find the value of the resistance of the resistor R.

   (d) Find the current in the 48 Ω resistor.

   (CCEA June 2009)

3. An electrician finds two coils of resistance wire in his bag. The coils of wire are made of materials of different electrical resistivity.

(a) Define electrical resistivity.

(b)(i) Coil A consists of wire 15 m long, with a diameter of 0.2 mm and a resistance of 9.0 Ω. Calculate the resistivity of the material of Coil A and state its unit.

(ii) Coil B, consists of wire of the same length and diameter as the wire in Coil A but with a resistivity 30 times that of Coil A. Calculate the resistance of Coil B.

(c) (i) The electrician has to fix two faults, a heating element and a break in the electrical circuit. He uses wire from the coils to mend the faults. Which coil of wire should the electrician select for each job?

(ii) Explain your choice of wire to repair the heating element.

(CCEA June 2010)

4. (a) A copper wire is 2.0 m long and has a radius of 0.56 mm. When the current in the wire is 3.5 A, the potential difference between the ends of the wire is 0.12V. Calculate the resistivity of copper.

(b) This copper wire (wire A) is now replaced with a different copper wire (wire B) of length 2.0 m (the same as before) but of radius 0.28 mm (half the previous value). State how the resistance and resistivity of wire B compare with the values of the corresponding quantities for wire A. In each case, explain your reasoning.

(CCEA January 2009)

5. The diagram shows a heating element as used in the rear window of a car. It consists of six strips of resistive material, joined by strips of copper of negligible resistance. The voltage applied to the heater is 14.2 V when the engine is running. The total current delivered to the heater by the battery is 8.4 A.

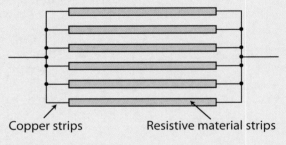

Copper strips   Resistive material strips

(a)(i) Calculate the total resistance of the element.
(ii) Calculate the power delivered by the battery to the heating element.

(b) Calculate the resistance of one of the strips of resistive material.

(c) The heating element has six strips connected in parallel. Suggest two reasons why this arrangement is preferable to connecting the same strips in series. (CCEA June 2009, amended)

6. The diagram shows an arrangement of resistors.
(a) Calculate the resistance between terminals X and Y.
(b) An additional 6 Ω resistor is connected between terminals X and Y so that it is in parallel with both pairs

of 6 Ω resistors. Calculate the total resistance between terminals Y and Z.

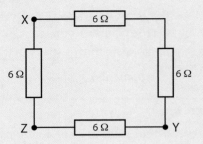

(CCEA June 2010, amended)

7. (a) State Ohm's law.

(b) An arrangement of four identical resistors is shown below. The potential difference between points X and Y is 6.0 V. The current entering at X is 2.0 mA.

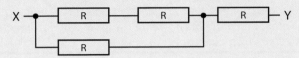

(i) What is the total resistance between points X and Y?

(ii) Calculate the resistance of one of the resistors.

(CCEA June 2016)

8. The diagram shows a number of resistors connected in series and parallel to a 15 V battery of negligible internal resistance. The overall resistance of the circuit is 12.5 Ω.

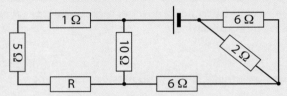

(a) Calculate the value of the resistor labelled R in the circuit.

(b) Calculate the current flowing through the 2 Ω resistor and hence the power dissipated in it.

(CCEA June 2015)

9. (a) Derive an equation for the resistivity of a sample of wire, in terms of resistance R, diameter d and length L of the wire.

(b) A student carried out an investigation into how the resistance of a metal wire varied with length of wire. She plotted values of length in metres against resistance in ohms and obtained the graph shown.

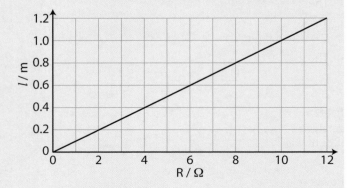

The student cannot recall whether the metal wire is a sample of aluminium (resistivity 2.82×10^{-8} Ω m) or nichrome (resistivity 1.00×10^{-6} Ω m). Use the equation you stated in (a) and data from the graph to carry out calculations to decide which of these two materials was used in the experiment, and explain your decision.

(c) State how the student may have ensured the investigation was carried out in as safe a manner as possible.

(CCEA June 2016, amended)

10. The fuse fitted to a three pin plug is designed to melt when the current exceeds 13 A. It is made of a piece of fuse wire, 25.4 mm long, of resistivity 1.45×10^{-6} Ωm and has a resistance of 0.19 Ω.

(a) What is the minimum diameter of fuse wire which must be used if it is to allow 13.0 A to pass through the fuse?

(b) If it was required that the fuse wire melted with a smaller current, how must the wire be changed if it has the same length and is made of the same material? Explain your answer.

(CCEA June 2014)

11. A certain thermistor's resistance at 20°C is 2860 Ω and at 100°C is 199 Ω. Explain why the resistance of the thermistor varies with temperature in the manner that it does.

(CCEA June 2013, amended)

1.11 Internal Resistance and Electromotive Force

Students should be able to:

1.11.1   demonstrate an understanding of the simple consequences of internal resistance of a source for external circuits;

1.11.2   use the equation $V = E - Ir$;

1.11.3   perform and describe an experiment to measure internal resistance and the electromotive force;

Internal Resistance

Sources of e.m.f. (electromotive force), such as batteries and power packs, have themselves some resistance to the electric current that passes through them. This is called their **internal** resistance and it has two effects:

1. As more current is drawn from the battery or power pack, the voltage across the terminals of the supply falls.

2. The source of e.m.f. is less than 100% efficient as energy is dissipated as heat within it.

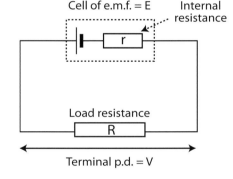

The e.m.f. is the **open-circuit voltage** (this is the p.d. across the terminals of the cell when no current is drawn from it). The internal resistance of a source of e.m.f. may be thought of as a **resistance, r, in series with the supply**. In the circuit on the right and in the discussion below:

E = e.m.f. of the cell, in V
V = voltage across the terminals, V
r = internal resistance, in Ω
R = load resistance, in Ω
v = voltage lost in internal resistance, in V

•   Total resistance = $R + r$

•   Current $I = \dfrac{E}{(R + r)}$

•   The potential difference across the load resistance is known as the **terminal potential difference**, V, where $V = IR$.

•   The potential difference lost in the internal resistance is $v = Ir$, so:
    $E = V + v = IR + Ir = I(R + r)$

•   The maximum current that can be taken from a power supply occurs when the cell is short-circuited (so there is no load resistance, $R = 0$) is $I_{max} = \dfrac{E}{R}$

Internal Resistance and the Law of Conservation of Energy

Analysis of the circuit above shows that: $E = V + Ir$
If both sides of the equation are multiplied by I, the result is:

EI	$=$	VI	$+$	I^2r
Power released by chemical energy in battery		Power delivered to the the external circuit		Power dissipated in the internal resistance of the battery

This is simply an application of the law of conservation of energy to a battery with an internal resistance.

Worked Example

When a 12 V battery is short-circuited, the current drawn is 6 A. What current would you expect to flow when the load resistor is 4 Ω?

Internal resistance = V ÷ I = 12 ÷ 6 = 2 Ω.
When the load is 4 Ω, the total resistance is 2 + 4 = 6 Ω, so I = V ÷ R = 12 ÷ 6 = 2 A.

Worked Example

Three identical cells each have an internal resistance of 0.5 Ω. They are connected in series with each other across a load resistor of 1.5 Ω. If the e.m.f. of each cell is 2.0 V, calculate the current drawn from the battery and the power dissipated in the load resistor.

I = battery voltage ÷ circuit resistance
  = (3 × 2.0) ÷ (3 × 0.5 + 1.5) = 2.0 A
Power in external resistor = I²R = 2.0² × 1.5 = 6.0 W

Experiment to find the Internal Resistance of a Cell

Determination of the internal resistance of a cell is an experiment prescribed by the specification. You should therefore be able to describe the experiment in detail.

The procedure is as follows:

- Using a 'D' cell (commonly called a torch battery) and a 5 Ω rheostat set to its highest resistance, assemble the circuit as shown.
- Record the terminal voltage, V, and current, I, drawn from the cell.
- Reduce the load resistance slightly and record the new values of V and I.
- Repeat for different load resistances, R ranging from 5 Ω to zero.

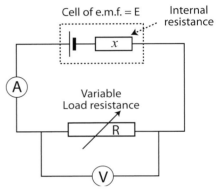

Theory

E = I(r + R) = Ir + IR = Ir + V

Rearranging gives: V = E – Ir

Comparing this with $x = mx + c$ shows that a graph of V against I gives a straight line of slope –r and y-axis intercept of E.

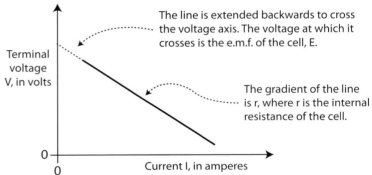

The line is extended backwards to cross the voltage axis. The voltage at which it crosses is the e.m.f. of the cell, E.

The gradient of the line is r, where r is the internal resistance of the cell.

Exercise 11

1. A source of electromotive force has an internal resistance.
   (a) Describe an effect of this internal resistance.
   (b) An experiment is performed to determine the internal resistance r of a cell using the circuit below.

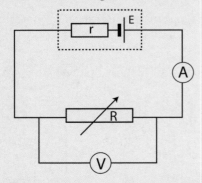

By varying the load resistance, R, results were obtained for the terminal voltage V and the current drawn from the cell. A graph of the results was plotted as shown.

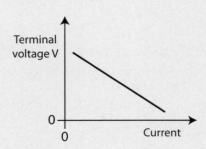

   (i) Explain how the internal resistance r may be obtained from the graph.
   (ii) Explain how the e.m.f. E of the cell may be obtained from the graph.

(c) When no current is drawn from the cell, the p.d. between its terminals is 10.0 V. When a load resistor of 2.0 Ω is connected across the battery the p.d. between the terminals is 9.5 V. Calculate the internal resistance of the cell.

(CCEA June 2010)

2. (a) Terms often used in describing an electrical power source, such as a battery, are terminal potential difference and electromotive force (e.m.f.). Write a short explanation of these terms and when it is appropriate to use them.

(b) (i) What is meant by the internal resistance of an electrical power source?

(ii) Describe an experiment to find the internal resistance of a battery. Include a circuit diagram. Show how a value of the internal resistance can be obtained from the series of experimental results. Show also how the e.m.f. of the battery can be obtained.

(CCEA June 2009)

3. Determining the internal resistance r of a cell requires a circuit to be set up that enables quantities to be measured that when analysed allow the internal resistance to be obtained.

(a) Draw the circuit diagram which will enable you to take readings from which you can determine the value of r.

(b) State the quantities to be used to plot a graph from which r can be determined. Label the axes, on the lines provided, and sketch the shape of graph expected.

(c) Explain how the value of the internal resistance of the cell is obtained from your graph.

(CCEA June 2013)

4. In the circuit shown a four cell battery is connected to a 16.4 Ω resistance through a switch. The table provides the meter readings with the switch open and closed.

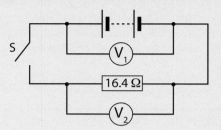

s	V_1	V_2
open	6.52 V	0.00 V
closed	5.33 V	5.33 V

(a) Determine the average internal resistance of a cell.

(b) If this circuit was set up in reality, it is unlikely that the voltmeters would read exactly the same value with the switch closed. Which reading would be higher and why?

(CCEA January 2014)

1.12 Potential Divider Circuits

Students should be able to:

1.12.1 demonstrate an understanding of the use of a potential divider to supply variable potential difference from a fixed power supply;

1.12.2 demonstrate knowledge and understanding of the use of the potential divider in lighting and heating control circuits; and

1.12.3 calculate the output voltages in loaded circuits using the equation $V_{out} = \dfrac{R_1 V_{in}}{R_1 + R_2}$.

Potential Dividers

A potential divider is an arrangement of resistors which allows a fraction of the p.d. supplied to it to be passed on to an external circuit.

In the circuit shown the current, I in resistor R_1 is given by:

$$I = \frac{V_{in}}{R_1 + R_2}$$

The output voltage is the potential difference across R_1:

$$V_{OUT} = I \times R_1$$

Applying Ohm's law to R_1:

$$V_{out} = \frac{R_1 V_{in}}{R_1 + R_2}$$

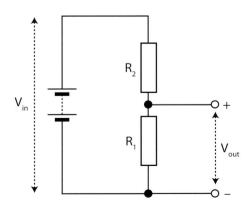

Potential divider with fixed resistors

41

With fixed resistors, the potential divider **does not permit the user to vary the output voltage**. To do that the fixed resistors are replaced by a **rheostat**, as shown below. A rheostat used in this way is called a **potentiometer**. The output voltage is then continuously variable. The output voltage is then connected across a **load resistor**.

Note that **the output voltage is simply a fraction of the input voltage**. Students must understand that the fraction involved is the ratio of the resistance across which the output voltage is taken to the total resistance of the circuit.

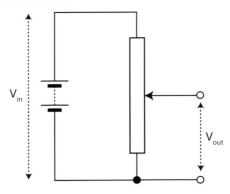

Potential divider with a rheostat

Worked Example

The circuit shows a potential divider.

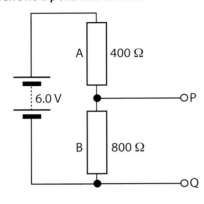

(a) What is the magnitude of the potential difference between the terminals P and Q?

(b) A resistor of resistance 400 Ω is now connected in parallel with the 800 Ω. Calculate the new p.d. between P and Q.

(a) $V_{OUT} = V_{PQ} = \dfrac{R_1}{R_1 + R_2} \times V_{IN} = \dfrac{800}{1200} \times 6 = 4.0\,V$

(b) The combined resistance of the resistors in parallel is

$R_T = \dfrac{R_1 \times R_2}{R_1 + R_2} = (800 \times 400) \div (800 + 400) = 266.67\,\Omega$

$V_{OUT} = V_{PQ} = \dfrac{R_1}{R_1 + R_2} \times V_{IN} = \dfrac{266.67}{(266.67 + 400)} \times 6.0 = 2.4\,V$

Effect of loading on V_OUT

In the worked example, the addition of the 400 Ω across the terminals P and Q has reduced the maximum output voltage of the potential divider from 4.0 V to 2.4 V. In general, **the voltage across the load decreases as the resistance of the load decreases**. Note that:

• Where there is infinite load, resistance V_{OUT} is a maximum.

• If there is zero load, resistance, V_{OUT} is zero.

Exercise 12

1. A 15 V battery is connected to a circuit that provides a voltage V_{out} that depends on the brightness of a room. The brightness of the room is sensed by a light-dependent resistor (LDR). This LDR is connected to a 300 Ω fixed resistor to form a potential divider as shown in **circuit 1**. The resistance of the LDR varies from a minimum of 10 Ω in bright conditions to a maximum of 250 Ω in dark conditions.

(a) Calculate the output voltage V_{out} when the room is brightly lit.

The circuit is altered so that an external load containing a small motor is connected across the LDR. This causes a window blind to be closed automatically when the room becomes dark. The voltage across the motor must be 6 V for the motor to close the blind. The altered circuit is shown in **circuit 2**.

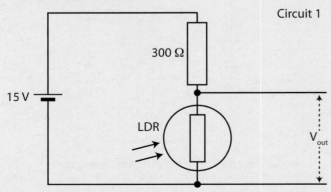

Circuit 1

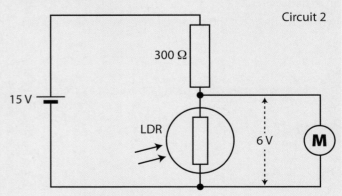

Circuit 2

(b) Calculate the resistance of the external load circuit containing the motor.

(CCEA June 2015)

2. The diagram shows a potential divider circuit. A voltmeter of resistance 20 kΩ is connected as shown. The values of R_1 and R_2 are both equal to 20 kΩ.

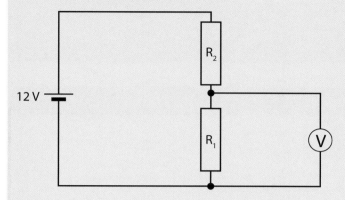

(a) Calculate the value of the reading on the voltmeter.

(b) The voltmeter is removed from the circuit and replaced by another voltmeter of resistance 10 MΩ. Explain why the output voltage is the same as when no voltmeter is connected across resistor R_1.

(CCEA June 2014)

3. Below is the circuit diagram for a potential divider that incorporates a thermistor and a variable resistor initially set at 3180 Ω. This potential divider circuit is to be used to control the temperature in an incubator. A heater will switch on when the potential difference across the thermistor is 6.0 V.

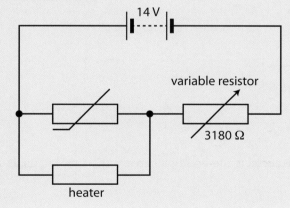

(a) Show that the thermistor resistance that produces a p.d. of 6.0 V across the thermistor is 2400 Ω (to 2 sig. figs).

(b) The heater element is part of a heating circuit that has to be placed in parallel with the thermistor. If the heating element has a resistance of 1600 Ω, calculate the new resistance to which the variable resistor must be adjusted if the heater's switch-on voltage is to remain at 6.0 V.

(CCEA June 2013 amended)

4. The diagram shows a circuit used to turn on a light automatically when it gets dark. It makes use of a light dependent resistor, LDR, the resistance of which depends on the amount of light shining on it and a fixed resistor of resistance 10 kΩ.

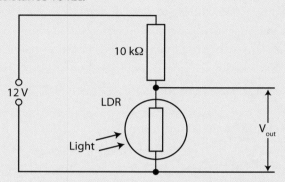

(a) The LDR has a resistance of 500 Ω in bright light and 200 kΩ when it is dark.

(i) Calculate the output voltage V_{out} when the LDR is in bright light.

(ii) The lamp connected across the output V_{out} lights when V_{out} is greater than 10 V. Show that the lamp will light in the dark.

(b) Describe and explain what effect swapping the positions of the LDR and the fixed resistor in the circuit would have.

(CCEA January 2010, amended)

Unit AS 2:
Waves, Photons and Astronomy

Students should be able to:

2.1.1 demonstrate knowledge and understanding of the terms transverse wave and longitudinal wave;

2.1.2 categorise waves as transverse or longitudinal;

2.1.3 analyse graphs to obtain data on amplitude, period, frequency, wavelength and phase;

2.1.4 demonstrate an understanding that polarisation is a phenomenon associated with transverse waves;

2.1.5 recall and use the equations $f = \frac{1}{T}$ and $v = f\lambda$;

2.1.6 recall radio waves, microwaves, infrared, visible, ultraviolet, X-rays and gamma rays as regions of the electromagnetic spectrum;

2.1.7 state typical wavelengths for each of these regions;

2.1.8 recall that the wavelength of violet light is 400 nm and red light is 700 nm;

Transverse and Longitudinal Waves

Waves are created by a disturbance which results in a vibration. **A wave that transports energy by causing vibrations in the material or medium through which it moves is called a progressive wave.**

In a **transverse wave** the vibrations are **perpendicular** to the direction in which the wave travels (ie the wave carries energy or propagates) as shown in the diagram. Electromagnetic waves, water waves and waves on ropes are all examples of longitudinal waves.

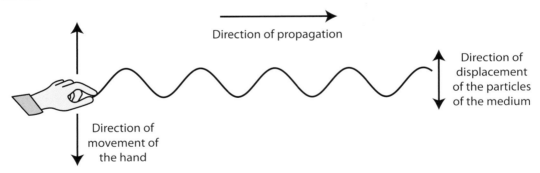

In a **longitudinal wave** the vibrations are **parallel** to the direction of propagation as shown in the diagram. Sound and ultrasound are examples of longitudinal waves.

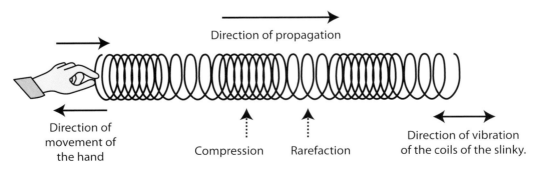

Definitions

The **period** of a wave, T, is the time for a particle in a medium to make one complete cycle.

The **frequency** is the number of complete waves that pass a point in one second, or the number of complete oscillations of a particle in the medium in one second. Frequency is measured in hertz, (Hz). A frequency of 100 Hz means 100 waves per second pass a point or the particle of the medium completes 100 oscillations in one second. Frequency is equal to $\frac{1}{T}$.

The **wavelength** is defined as the distance the wave form progresses in the periodic time, T. The wavelength can be measured as the distance from crest to next crest, or from trough to next trough.

The **amplitude** of a wave is the maximum displacement of a particle of the medium from its rest position.

Graphical Representation of a Wave

Both transverse and longitudinal waves can be represented graphically in two ways:
1. Displacement of a particle of the medium against time.
2. Displacement of the particles of the medium against distance along the wave.

This graph shows how the displacement, from its equilibrium position, of a particle of the medium through which the wave is moving, varies with time.

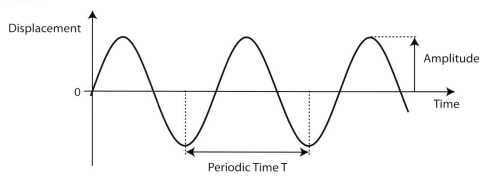

This graph shows how the displacement, from their equilibrium position, of particles of the medium through which the wave is moving, varies with distance along the direction in which the wave is travelling.

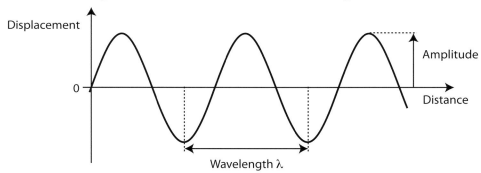

It is **important to remember** that you obtain the **period** (and hence the **frequency**) from a **displacement-time graph**. You obtain the **wavelength** from a **displacement-distance graph**.

Phase

The particles of the medium through which a wave passes vibrate. If two particles are vibrating so that at the same instant they are at the same distance and same direction (ie, the same displacement) from their equilibrium positions, they are said to be **in phase**. Phase is also used to describe the relative positions of crests and troughs on two waves of **the same frequency**. If the crests of one wave coincides with the crests of the other, we say the waves are in phase. If the crests of one wave coincide with the troughs of the other, we describe them as being out of phase by ½λ.

Phase can be expressed in four ways:
• a fraction of a wavelength
• a fraction of a period
• an angle in degrees
• an angle in radians

To find the phase difference in degrees:

$$\text{phase difference} = \frac{x \times 360°}{\lambda} \text{ or } \frac{t \times 360°}{T}$$

To find the phase difference in radians:

$$\text{phase difference} = \frac{x \times 2\pi}{\lambda} \text{ or } \frac{t \times 2\pi}{T}$$

where x is the distance between the peaks of the two waves and
                 t is the time interval between the peaks.

A full wavelength or a full period represents an angle of 360° or 2π radians.

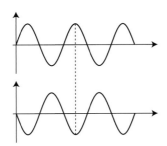 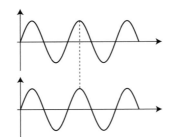 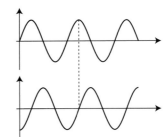

These two waves are out of phase. The crests of one exactly coincide with the troughs of the other. Their phase difference is ½ λ.

These two waves are in phase. The crests of one exactly coincide with the crests of the other.

These two waves are out of phase. The crests of one exactly coincide with the point where the displacement of the other wave is zero. Their phase difference is ¼ λ.

Worked Example

Consider two transverse waves A and B represented by the graphs below. What is the phase difference between the waves?

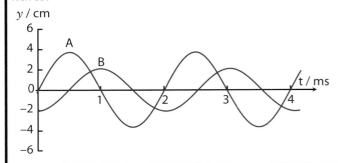

The period of each wave is 2 ms (from the graph).

Wave A reaches its peak after 0.5 ms, wave B reaches its peak after 1.0 ms.

So the time interval between the peaks is 0.5 ms.

But this time interval represents ¼ × the period or $\frac{T}{4}$.

So the phase difference is ¼ × 360° = 90° or $\frac{\pi}{2}$ radians.

Polarisation

An electromagnetic wave is a transverse wave which has both an electric and a magnetic oscillating component. A light wave which is vibrating in more than one plane is referred to as unpolarised light. Polarised light waves are light waves in which the vibrations occur in a single plane. The process of transforming unpolarised light into polarised light is known as polarisation and can be achieved using a polarising filter.

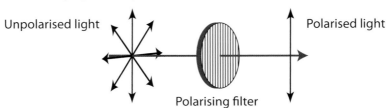

When you view the polarised light through a second polarising filter (analyser) the intensity of the transmitted light falls to a minimum after the analyser has been rotated through 90°. A further rotation of 90° will see the transmitted light intensity reach a maximum again.

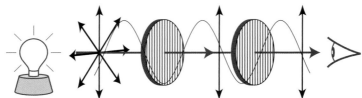

Velocity of a Wave

The velocity of a wave can be calculated from its wavelength and frequency using the following equation:

$v = f\lambda$ where v = velocity in m s^{-1}
f = frequency in Hz
λ = wavelength in m

Spectrum of Electromagnetic Waves

Electromagnetic waves exist with a very large range of wavelengths. This continuous range of wavelengths is known as the electromagnetic spectrum:

Wave	Gamma	X ray	Ultraviolet	Visible	Infrared	Microwave	Radio
Wavelength/m	$\leq 10^{-11}$	10^{-10}	10^{-8}	Violet = 4×10^{-7} Red = 7×10^{-7}	10^{-4}	10^{-2}	~1
Frequency/Hz	$\geq 3 \times 10^{19}$	3×10^{18}	3×10^{16}	Violet = 7×10^{14} Red = 4×10^{14}	3×10^{12}	3×10^{10}	~3×10^8

Students should note that they are now required to know that the range of visible light is from 400 nm to 700 nm.

Worked Example

(a) State a typical wavelength for visible light.

(b) An electromagnetic wave from a different region of the electromagnetic spectrum has frequency of 620 GHz. What is its wavelength of it is travelling in a vacuum?

(CCEA June 2009)

(a) Visible light has wavelengths from 400 nm (4×10^{-7} m) (violet) to 700 nm (7×10^{-7} m) (red).

(b) Use the wave equation $v = f\lambda$.
f = 620 GHz = 620×10^9 Hz.
$v = 3 \times 10^8$ m s^{-1} (from data sheet)
Re-arranging gives $\lambda = 3 \times 10^8 \div 620 \times 10^9 = 4.84 \times 10^{-4}$ m

---

Worked Example

The graphs of displacement against time for two waves A and B are shown opposite.

(a) (i) State the amplitude of A.
(ii) Calculate the frequency of wave A.

(b) Are the graphs shown useful in classifying the waves as transverse? Explain your answer.

(c) (i) Waves A and B are not in phase. Explain what is meant by 'in phase'.
(ii) What is the phase difference between wave A and wave B?
(CCEA January 2010)

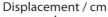
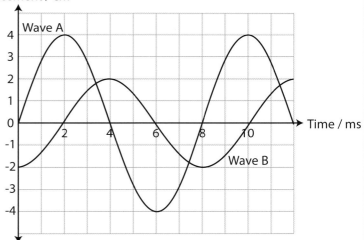

(a) (i) The amplitude is the maximum displacement of produced by the wave. Unlike the displacement it is not a vector so + or – is not required. Reading from the graph, making sure you note the scale on the vertical axis, it is 4 cm.
(ii) The frequency = 1/period. The horizontal axis will give the time for one complete wave. Reading from the graph, noting the scale is in ms (milliseconds, 10^{-3} s), gives 8 ms or 8×10^{-3}. Frequency = $1 \div 8 \times 10^{-3}$ = 125 Hz.

(b) No. Although the graph shows displacement against time, it does not show if the vibrations are perpendicular to the direction of propagation of the wave (transverse) or parallel to the direction of propagation (longitudinal).

(c) (i) Corresponding points on each wave coincide in time. Or one wave reaches its maximum positive displacement at the same time as the other wave also reaches its maximum positive displacement.
(ii) Wave A reaches its maximum positive displacement 2 divisions before wave B reaches its maximum positive displacement. 2 divisions = 2 ms. This is ¼ of the period, so the phase difference is ¼ T or ¼ wave = ¼ of 360° = 90° or ¼ of $2\pi = \pi/2$ radians.

Exercise 13

1. (a) Name the regions of the electromagnetic spectrum in order of decreasing frequency.

    (b) State the approximate limits of the visible part of the electromagnetic spectrum.

    (c) In every transverse wave there is an oscillation perpendicular to the direction in which the wave is moving. State what is oscillating in an electromagnetic wave and illustrate your answer with a suitably labelled diagram.

    (d) State a property unique to electromagnetic waves.

2. A wave of amplitude 10 cm and frequency 5 Hz travels along a stretched wire at speed of 20 m s⁻¹.

    (a) Calculate the period and sketch a graph of displacement against time to represent the wave.

    (b) Calculate the wavelength and sketch a graph of displacement against distance to represent the wave.

    (c) Calculate the phase difference between two points on the wire which are 0.8 m apart, giving your answer in degrees.

3. State the period and frequency of the wave shown below.

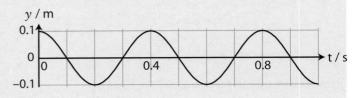

4. (a) Sketch graphs of:

    (i) wavelength (in m) against frequency (in Hz) for sound waves in air.

    (ii) speed (in m s⁻¹) against wavelength (in m) for electromagnetic waves in air.

    (b) A graph is plotted of frequency (in Hz) against λ^{-1} (in m⁻¹) for light waves. Sketch the graph and find the value of its gradient. Remember the gradient has a unit.

2.2 Refraction

Students should be able to:

2.2.1 perform and describe an experiment to verify Snell's law and measure the refractive index;

2.2.2 recall and use the equations $n = \dfrac{\sin i}{\sin r} = \dfrac{c_1}{c_2}$;

2.2.3 demonstrate knowledge and understanding of total internal reflection; and

2.2.4 recall and use the equation $\sin c = \dfrac{1}{n}$;

2.2.5 demonstrate an understanding of the physical principle of the step index optical fibre, including total internal reflection at the core/cladding boundary and the speed in the core;

2.2.6 describe the structure of a flexible endoscope and discuss examples of its application in medicine and industry;

Snell's Law

Refraction occurs when a wave (for example light) travels from one medium to another (for example from air into glass). When a light ray is refracted, its direction of travel is changed. The angle between the incident ray and the normal is called the **angle of incidence**, i. The angle between the normal and the refracted ray is called the **angle of refraction**, r. Note that the angles are measured from the normal.

Snell's Law states that for light travelling from one material to another, the ratio $\dfrac{\sin i}{\sin r}$ is a constant known as the refractive index. The refractive index of a material, n, can also be calculated

from: $n = \dfrac{\text{speed of light in a vacuum (or air)}}{\text{speed of light in the material}}$.

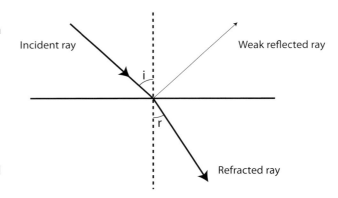

Experimental Verification of Snell's Law and Measurement of Refractive Index

Snell's Law can be verified by ray tracing through a glass block. Place the glass block on a sheet of paper and carefully trace around it. Remove the glass block and mark the normal at one edge and extend this line into the position of the glass block. Replace the glass block.

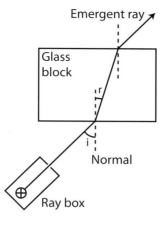

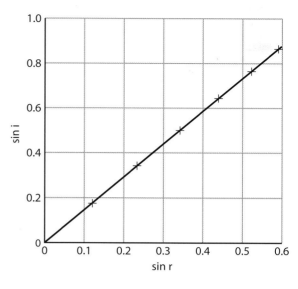

A ray box is used to produce a narrow ray. Shine the ray into the block so that it meet the block at the point where the normal meets the block. Mark this path carefully with crosses. Mark the emergent ray in a similar fashion.

Remove the glass block, join up the crosses to show the incident, refracted and emergent rays. Using a protractor measure the angles of incidence and refraction.

Carefully replace the glass block and repeat this procedure for a number of incident rays with angles of incidence. When sin i is plotted against sin r, the result is a straight line through the origin, verifying Snell's Law. The gradient of the line gives the value for the refractive index.

Critical Angle

When light travels from a material of high refractive index to one of lower refractive index, for example from glass into air, it is bent away from the normal. The angle of incidence in the glass that produces this angle of refraction of 90° is called the **critical angle C**. Note that there is still a weak reflected ray.

The relationship between the critical angle and the refractive index of the material can be derived by applying Snell's Law:

Since the refracted ray is at 90° to the normal:

$$_{glass}n_{air} = \frac{\sin C}{\sin 90} = \sin C$$

$$_{glass}n_{air} = \frac{1}{_{air}n_{glass}}$$

$$\sin C = \frac{1}{_{air}n_{glass}}$$

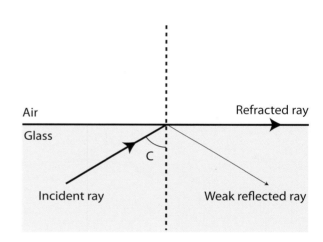

Total Internal Reflection

When the angle of incidence in the material with the higher refractive index is greater than the critical angle, a phenomenon known as **total internal reflection** occurs. Total internal reflection involves the reflection of all the incident light at the boundary between two materials.

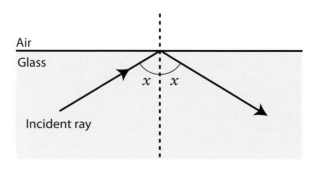

Worked Example

The diagram shows the outline ABCD of a rectangular block made of glass of refractive index 1.46. A ray of light is incident at an angle of incidence of 75.0° on side AB. The angle of refraction for this ray is 41.4°. The refracted ray meets face AD.

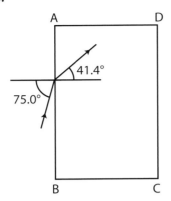

(a) Calculate the angle of incidence of this ray when it meets face AD.

(b) Calculate the critical angle for a ray in this glass meeting the glass/air boundary.

(c) Use your answers to deduce what will happen to the ray when it meets face AD.

(d) On the diagram continue the path of the ray until it has left the glass block.

(CCEA January 2010)

(a) From the geometry (see diagram) the angle is
90° − 41.4° = 48.6°.

(b) sin c = 1 ÷ n = 1 ÷ 1.46 = 0.6849 giving c = 43.2°.

(c) The ray makes an angle of incidence of 48.6° with the side AD. This angle is greater than the critical angle resulting in total internal reflection at AD. The ray then meets the other side of the block at an angle of incidence of 41.4° and ray emerges into the air making an angle of refraction of 75°.

(d)

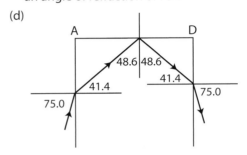

Optical fibres

The use of a long strand of glass to send light from one end of the medium to the other is the basis for modern day use of **optical fibres**. Optical fibres are used in communication systems and micro-surgery. On each occasion when the ray of light meets the glass/air boundary the angle of incidence exceeds the critical angle, and total internal reflection takes place. None of the incident energy is ever lost due to the transmission of light across the boundary. The intensity of the signal remains constant.

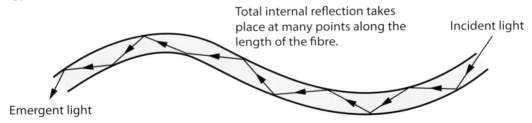

Total internal reflection takes place at many points along the length of the fibre.

Incident light

Emergent light

The step index fibre

The step index fibre consists of a glass core, typically 100 μm in diameter, surrounded by a glass cladding, typically 150 μm in diameter. Surrounding the core and cladding is a layer of protective plastic called a sheath, which is usually omitted from diagrams. **The refractive index of the core is slightly greater than that of the cladding.** This type of optical fibre is called 'step index' because the refractive index does not change gradually at the core/cladding boundary; rather it changes like a step. Provided the angle of incidence at the core/cladding boundary is greater than the critical angle, the light signal is propagated in the core by repeated total internal reflection.

The path of light along the centre of the core is called **axial mode**. The path of light that repeatedly meets the core/cladding boundary at the critical angle is called the highest order mode. There is clearly a difference between the time taken for light to travel up a fibre along the central axis and that travelling by repeated total internal reflections at just above the critical angle. This is called modal dispersion and is very undesirable. In an endoscope, modal dispersion would result in a blurred image of the target organ. Modal dispersion can be prevented by using **very thin cores** so that only axial mode is possible.

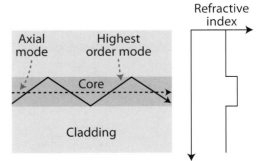

Worked Example

An optical fibre has a core of refractive index of 1.60 and a cladding of refractive index 1.50.

(a) Calculate the critical angle at the core/cladding boundary in this fibre.

(b) Calculate the ratio of the path length for the highest order mode to the axial mode in this fibre.

(c) What is the time difference between a light signal travelling along the axis of a 1 km length of this fibre and that of a signal travelling in the highest order mode?

(a) $_{air}n_{cladding} \times {}_{cladding}n_{core} \times {}_{core}n_{air} = 1$

$1.5 \times {}_{cladding}n_{core} \times {}^1/_{1.6} = 1$

$_{cladding}n_{core} = {}^{1.6}/_{1.5} = 1.067$

Critical angle $C = \sin^{-1}\left(\dfrac{1}{{}_{cladding}n_{core}}\right)$

$= \sin^{-1}\left(\dfrac{1}{1.067}\right) = 69.6°$

(b) Suppose the axial mode distance corresponding to a single total internal reflection is x.

Then the corresponding highest order mode distance is $\left(\dfrac{x}{\sin C}\right)$.

So the required ratio is $\dfrac{1}{\sin C}$ or $\dfrac{1}{\sin 69.6} = 1.07$

(c) Speed of light in the core $= \dfrac{\text{speed in air}}{{}_{air}n_{core}}$

$= \dfrac{3 \times 10^8}{1.6} = 1.88 \times 10^8 \text{ m s}^{-1}$

Distance travelled in axial mode $= 1000$ m

Time for axial mode signal $= \dfrac{1000}{1.875 \times 10^8} = 5.33$ μs

Distance travelled in highest order mode

$= 1000 \times 1.07 = 1070$ m

Time for highest order mode signal

$= \dfrac{1070}{1.875 \times 10^8} = 5.71$ μs

Time difference $= 0.37$ μs

Note: the most common mistake in part (c) is failing to find the speed of light in the core.

Optical fibres in medicine – the flexible endoscope

An endoscope is a flexible tube that allows us to look into the body. In many cases there is no need to perform surgery to do this. In other cases, a small incision is required to perform what has become known as key-hole surgery.

The endoscope has two bundles of optical fibres. One is called the **illumination bundle** and carries light to the object being viewed. The illumination bundle is described as **incoherent** since the fibres are randomly arranged. The other bundle, the **image bundle**, carries back the reflected light. The optical fibres inside the image bundle are carefully arranged parallel to each other to create what is termed a **coherent bundle**. The coherent bundle contains up to 10 000 individual optical fibres.

The image is viewed or photographed through a magnifying eyepiece. In some instances a TV camera is attached and the image displayed on a monitor.

When inserted through the mouth, the endoscope can be used to view the gastrointestinal tract (oesophagus, stomach and duodenum). Alternatively, in the technique called laparoscopy, the endoscope is inserted through the wall of the abdomen to study the liver, spleen and other organs. The information obtained in this way provides direct and often very clear evidence of bleeding ulcers, constrictions, benign and malignant tumours and cirrhosis of the liver.

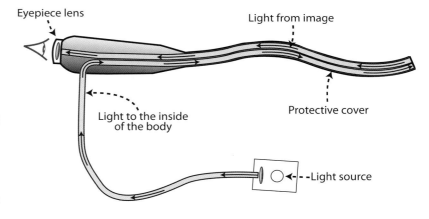

The endoscope also allows a range of minor surgical treatments. **Forceps** attached at the viewing end allows a surgeon to **remove a sample of tissue** (biopsy) for detailed analysis. Electrodes can be used to apply heat to stop bleeding. A range of **extractors** can be fitted and used to remove foreign objects from the throat or possibly drugs hidden in the lower bowel of smugglers. There is also a **water channel** to wash away the mucus from the end of the endoscope within the body.

Optical fibres in industry

The principles of industrial and medical endoscopes are similar. The differences are the applications and the associated tools. Industrial endoscopes are usually called borescopes and they are often used to view what is happening in inaccessible sites. A brief list of the applications is shown in the table below.

Industry	Application
Car	Inspection and repair of cylinders, exhaust pipes, manifolds, etc
Aircraft	Inspection and repair of engines, shafts, landing gear, turbines, etc
Ships	Inspection and repair of engines, turbines, heat exchangers, pipe lines, etc
Electricity generation	Inspection for cracks on mechanical equipment, boilers, turbines, pipes, etc
Espionage	Secretly viewing what is happening in another room
General	Internal check of pipes, tanks, steel pipes, etc

Exercise 14

1. In an experiment to measure the refractive index of glass, a series of results for a range of angles of incidence and their corresponding angles of refraction has been obtained. Describe how these results may be processed to obtain an accurate value for the refractive index of glass.
(CCEA June 2010)

2. A ray of light enters a medium of refractive index 1.39 at an angle as shown. The ray is refracted inside the medium and travels to the upper surface where it is incident at the critical angle.

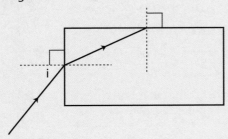

(a) Describe what happens to the ray at the upper surface.

(b) What would occur if another ray met the upper surface at an angle greater than the critical angle?

(c) Calculate the critical angle of the medium.

(d) Calculate the magnitude of the incident angle.

(CCEA January 2009)

3. A ray of light travels from inside a block of transparent material to air. The refractive index of the material of the block is 1.38. The ray emerges from the block into the air at angle of 43.0° to the normal. Calculate the minimum increase in the angle of the ray inside the block, to cause the ray to undergo total internal reflection.
(CCEA June 2010)

4. A short pulse of light enters a straight optical fibre of length 1.20 km. The pulse travels along the axis of the fibre as shown below.

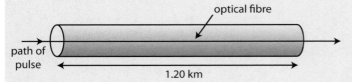

(a) The pulse takes 5880 ns to pass along the fibre. Calculate the velocity of light in the material of the fibre.

(b) Calculate the refractive index of the material.

(c) Calculate the critical angle of the material.

(CCEA January 2009)

5. (a) What is a step-index optical fibre?

The refractive index of the core of an optical fibre is 1.6 and the critical angle at the core-cladding boundary is 60°.

(b) Calculate the refractive index of the core-cladding boundary with respect to the air.

(c) Calculate the time difference between a light signal travelling along the axis of a 500 m fibre and that of a signal travelling in the highest order mode?

(d) An engineer is seeking to minimise the time delay calculated in (c) by changing the cladding which surrounds the core.

   (i) In what way, if any, would the refractive index of the cladding change?

   (ii) What effect, if any, would this have on the critical angle at the core-cladding boundary

   (iii) Carefully explain the advantage of using cores of a smaller diameter.

6. The diagram shows a prism, made of glass of refractive index 1.52. Each of the angles of the prism is 60.0°. A ray of light strikes the face AB at point P at an angle of incidence of 45.0°. After refraction at P, the ray crosses the prism to Q, where it is refracted out of the prism.

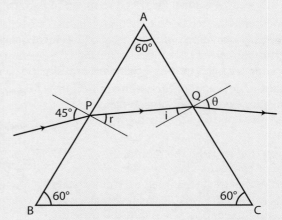

(a) Calculate the angle of refraction r of the ray after it enters the prism at P.

(b) The refracted ray crosses the prism, striking face AC at Q. Calculate the angle of incidence i of this ray on face AC. Hint: angle r + angle i = the angle of the prism.

(c) One of the conditions for total internal reflection is **not** satisfied when the ray meets face AC, so that the ray **does** emerge from the prism. State which condition is **not** met and support this with appropriate calculations.

(d) Calculate the angle θ to the normal at which the ray emerges from the prism.

(CCEA January 2008)

2.3 Lenses

Students should be able to:

2.3.1 draw ray diagrams for converging and diverging lenses;

2.3.2 use the equation $\dfrac{1}{u} + \dfrac{1}{v} = \dfrac{1}{f}$ for converging and diverging lenses;

2.3.3 verify experimentally the lens equation and the evaluation of f, the focal length of a converging lens, for real images only;

2.3.4 define m as the ratio of the image height to the object height, or $m = \dfrac{h_i}{h_o}$;

2.3.5 recall and use the equation $m = \dfrac{v}{u}$;

2.3.6 describe the use of lenses to correct myopia and hypermetropia;

2.3.7 perform calculations on the correction of long and short sight, including a calculation of the new range of vision; and

2.3.8 perform calculations involving the power of lenses.

Types of Lenses

Lenses can be classified into converging (convex) and diverging (concave). These two types of lens have a different effect on a parallel beam of light.

When parallel rays of light pass through a **converging** lens they are refracted so that they pass through the focal point or principal focus of the lens. This type of principal focus is described as **real**.

A **diverging** lens refracts the parallel rays so that they appear to spread out (diverge) from the focal point or principal focus of the concave lens. This type of principal focus is described as **virtual**.

The distance from the centre of a lens to the focal point is the focal length f.

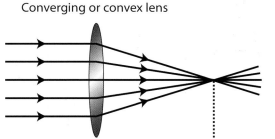

Converging or convex lens

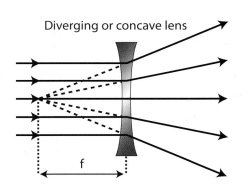

Diverging or concave lens

Ray Diagrams

The diagram below shows what happens to three particular rays when they pass through a **convex** lens. To find the position, size and nature of the image formed by a convex lens we need to find where at least two rays of light meet (real image) or appear to meet (virtual image) having passed through the lens from the from the object.

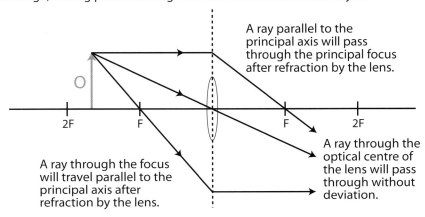

A ray parallel to the principal axis will pass through the principal focus after refraction by the lens.

A ray through the optical centre of the lens will pass through without deviation.

A ray through the focus will travel parallel to the principal axis after refraction by the lens.

The diagram below shows what happens to two particular rays when they pass through a **concave** lens. *Regardless of the position of the object*, the image formed is always virtual, erect and diminished. To locate the image we need to find the point where the two rays appear to meet (virtual image).

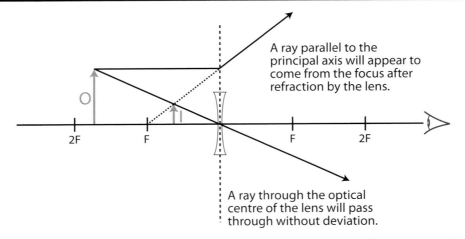

A ray parallel to the principal axis will appear to come from the focus after refraction by the lens.

A ray through the optical centre of the lens will pass through without deviation.

The table below summarises the location and properties of the image in a **convex** lens for various positions of the object.

Position of object	Position of image	Nature of the Image		
		Real / virtual	Enlarged /diminished	Upright /inverted
Between F and the lens	Further from the lens than the object and on the same side of the lens	Virtual	Enlarged	Upright
At F	At ∞	Real	Enlarged	Inverted
Between F and 2F	Beyond 2F	Real	Enlarged	Inverted
At 2F	At 2F	Real	Same size as the object	Inverted
Beyond 2F	Between F and 2F	Real	Diminished	Inverted
At ∞	At F	Real	Diminished	Inverted

Measurement of Focal Length (Converging lens)

There are three ways of measuring the focal length of a converging lens. Particular attention should be paid to the Plane Mirror Method and the Measuring Object and Image Distances Method.

Approximate Method

Light from a distant object, more than 10 m away, is approximately parallel. Adjust the distance from the lens to the screen so that image is as sharp as possible. Measure this distance; it is approximately the focal length. This should be repeated a number of times and take the average.

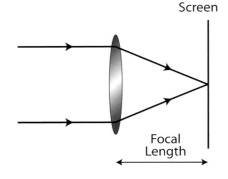

Plane Mirror Method

Place the plane mirror as close as possible to the lens. Move the lens and plane mirror together until a sharp image of the wire mesh appears on the front of the lamp house. Measure the distance from the lens to the lamp house, this is the focal length of the lens. Repetition and the taking of an average improves the reliability of the result.

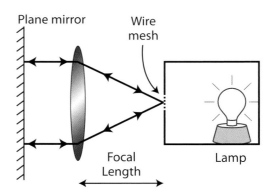

Measuring Object and Image Distances

Measure the distance from the mesh to the lens; this is the object distance u. The position of the screen is adjusted until a sharp image is produced on the screen. The lens formula can then be used to calculate the focal length:

$$\frac{1}{u} + \frac{1}{v} = \frac{1}{f}$$

The better approach is to obtain a series of values of u and v and use your results to plot a linear graph. This is achieved by plotting 1/u against 1/v. This yields a straight line as shown on the right. The intercept on each axes provides a value for 1/f.

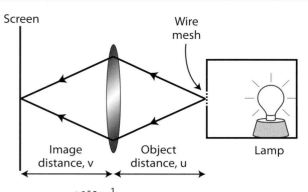

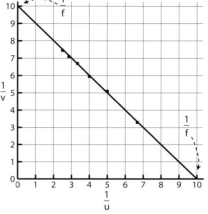

Verifying the Lens Equation

To verify the lens equation you should first find the focal length (using, say, the plane mirror method, or from data provided by the supplier) then carry out the experiment described above using object and image distances.

The final steps are:

- plot the graph of 1/v against 1/u and show it is a straight line;
- find the gradient and show that it is equal to –1; and finally
- confirm that the reciprocal of the intercept on each axis is equal to the value of f found by the plane mirror method.

Magnification

The linear magnification, m, of an image is the size of the image divided by the size of the object. By using the properties of similar triangles we can show that magnification is also equal to the ratio of the image distance to the object distance:

$$m = \frac{v}{u}$$

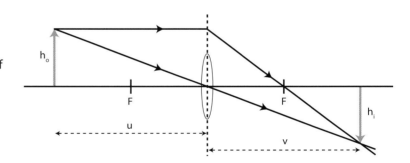

Power of a Lens

The power of lens, P, is defined as

$$P = \frac{1}{f}$$ where   f = focal length, in m
P = power, in m^{-1} or dioptres, symbol D

A diverging lens has a negative power, a converging lens has a positive power.

Real/Virtual Sign Convention

The distance to a real image is positive. The distance to a virtual image is negative.
A convex lens has a positive focal length. A concave lens has a negative focal length.

Worked Example

The lens of a projector produces an image of a square slide of side 2.00 cm on a screen 2.40 m from the lens. The image on the screen is a square of side 0.80 m.
(a) Describe the nature of the image formed on the screen.
(b) Calculate the distance between the projector lens and the slide.
(c) Hence calculate the focal length of the projector lens.
(CCEA January 2009)

(a) The image is real and magnified. It is also inverted and laterally inverted. (To ensure that the image is seen the correct way up and not laterally inverted the slide is actually placed in the projector upside down and also

laterally inverted.)
(b) The magnification m = image size ÷ object size
= 0.80 ÷ 0.02 = 40. The magnification is also equal to image distance ÷ object distance.
So the object distance is the distance from the lens to the slide = 2.40 ÷ 40 = 0.06 m or 6 cm
(c) The lens formula is used to find the focal length
$$\frac{1}{u} + \frac{1}{v} = \frac{1}{f}, \ \frac{1}{f} = \frac{1}{0.06} + \frac{1}{2.40} = 16.66 + 0.42 = 17.1$$

$$f = \frac{1}{17.1} = 0.059 \text{ m} = 5.9 \text{ cm}$$

Myopia and Hypermetropia

The ability of the eye to see objects clearly at different distances is known as **accommodation**.

The farthest point which can be seen clearly by the unaided eye is called the **far point**. For the normal eye this is at infinity. Light from the far point reaches the eye as parallel rays. The rays are refracted by the eye so that they meet on the retina forming a sharp image of the distant object.

The nearest point which can be seen clearly by the unaided eye is called the **near point**. For the normal eye this is at 25 cm. The light from the near point reaches the eye as diverging rays. These are refracted by the eye so that they meet on the retina forming a sharp image of the object at the near point.

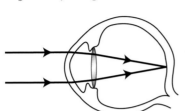

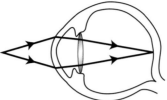

Myopia (Short Sight)

A person who suffers from myopia (short sight) is unable to see distant objects sharply. This causes the light from distant objects to converge towards a point in front of the retina. The image seen by the person is blurred. The person's far point is much closer to the eye than the normal infinite distance.

To correct this defect a concave (diverging) lens is used. **The focal length of the lens is equal to the distance to the person's actual far point.** This means that parallel rays of light from a distant object are refracted so that they appear to diverge from the person's far point.

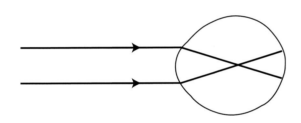

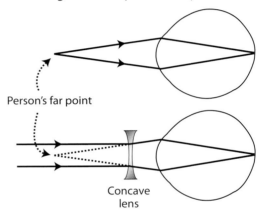

Hypermetropia (Long Sight)

A long sighted person sees distant objects clearly but does not see near objects clearly. An object held at the normal near point distance of 25 cm will not be seen clearly. The rays of light from the object are not bent sufficiently to form an image on the retina.

The near point is much further than 25 cm. Rays of light from an object placed at their near point are bent so that they meet on the retina resulting in the object being seen clearly. To correct for this defect a convex lens is used. **The focal length of this lens has to be such that an object at 25 cm appears to be at the person's near point**.

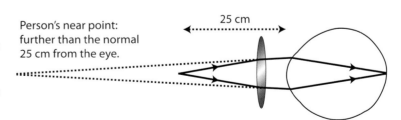

Worked Example

A person has a near point at 100 cm. Calculate the focal length and the power of the lens needed to correct this defect.

The convex lens has to create a **virtual image** at 100 cm of an object at 25 cm. Because the image at the person's near point is virtual, we use a negative sign.

If the person's near point is 100 cm from his eye then:

$$\frac{1}{u} + \frac{1}{v} = \frac{1}{f} \qquad \frac{1}{25} + \frac{1}{(-100)} = \frac{1}{f} \text{ so } \frac{1}{f} = \frac{3}{100}$$

giving f = 33.3 cm (0.333 m)

or a power = $\dfrac{1}{0.333}$ = +3.0 D

The convex lens needed has a focal length of 33.3 (0.333 m) and a power of +3.0 D

The current specification requires students not only to perform calculations on the correction of long sight and short sight, but also **to calculate of the new range of vision when using the correcting lens**. This is illustrated in the example below.

Worked Example

The range of vision of a patient is 40 cm – 400 cm. If the patient wears a lens of power –0.2 D, what is the range of vision now?

The patient's near point is 40 cm. We must first find the location of a real object that will give a virtual image at 40 cm with this lens.

Power $P = -0.2$ D, $f = \dfrac{1}{P} = \dfrac{1}{(-0.2)} = -5$ m $= -500$ cm

Using $\dfrac{1}{u} + \dfrac{1}{v} = \dfrac{1}{f}$: $\dfrac{1}{u} + \dfrac{1}{(-40)} = \dfrac{1}{(-500)}$, so $\dfrac{1}{u} = \dfrac{23}{1000}$

giving $u = 43.5$ cm.

The patient's far point is 400 cm. We must now find the location of the real object that will give a virtual image at 400 cm.

Using $\dfrac{1}{u} + \dfrac{1}{v} = \dfrac{1}{f}$: $\dfrac{1}{u} + \dfrac{1}{(-400)} = \dfrac{1}{(-500)}$, so $\dfrac{1}{u} = \dfrac{1}{2000}$

giving $u = 2000$ cm $= 20$ m

So with the correcting diverging lens, the range of vision is 43.5 cm to 20 m.

Exercise 15

1. An object labelled O is placed in front of a diverging lens L as shown.

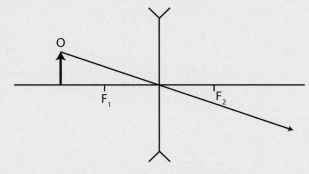

   Copy and complete the ray diagram to locate the position if the image obtained. Label the image I. The principal axis of the lens and the ray incident on the optical centre of the lens are included in the diagram. The locations of the principal foci are marked F_1 and F_2.

   (CCEA June 2011)

2. Describe an experiment to determine the focal length of a converging lens. Your description should include
   (a) a fully labelled diagram of the apparatus you intend to use,
   (b) an outline of the method employed,
   (c) the results to be taken,
   (d) an analysis of how the results can be used to obtain a value for the focal length of the converging lens.

   The quality of written communication was assessed in this question.

   (CCEA January 2010)

3. The focal length of a converging lens used to form a virtual and magnified image is 10.0 cm.
   (a) Calculate how far away from the lens an object must be placed in order to form a virtual image which is six times larger than the object.
   (b) Calculate how far from the lens this image is formed.
   (CCEA June 2008)

4. With the help of labelled diagrams describe and explain what is meant by:
   (a) short sight;
   (b) long sight.

5. A certain student requires spectacles in order to see a book clearly at normal reading distance.
   (a) What defect of vision do the spectacle lenses correct?
   (b) What is the normal cause of this defect?
   (c) Without spectacles, the student cannot see clearly objects which are nearer than 80 cm from the eye. Calculate the focal length of the spectacle lens required to reduce this distance to the normal near-point distance of 25 cm.
   (d) Without spectacles, the student's far-point distance has the normal value.
      (i) Calculate the distance of the student's far point from the eye when the spectacles in (c) are worn.
      (ii) When wearing the spectacles in (c), the student looks at a road sign which is 5 m away. Comment briefly on the appearance of the sign to the student.

6. The lens of the human eye is said to be unaccommodated when viewing objects at an infinite distance. The lens of a normal unacccommodated eye has a power of +50 D. Calculate:
   (a) the distance between the lens and the retina;
   (b) the increase in power of the lens when it focuses on an object at the normal near point.

2.4 Superposition, Interference and Diffraction

Students should be able to:

2.4.1 illustrate the concept of superposition by the graphical addition of two sinusoidal waves;

2.4.2 demonstrate an understanding of the conditions required to produce standing waves;

2.4.3 demonstrate knowledge and understanding of the graphical representation of standing waves in stretched strings, and air in pipes closed at one end;

2.4.4 identify, graphically, the modes of vibration of stretched strings and air in a pipe closed at one end, without reference to overtone and harmonic terminology;

2.4.5 identify node and antinode positions;

2.4.6 perform and describe an experiment to measure the speed of sound in air using a resonance tube (end correction is not required);

2.4.7 demonstrate an understanding of the conditions for observable interference;

2.4.8 demonstrate an understanding of the significance of path difference and phase difference in explaining interference effects;

2.4.9 describe Young's slits interference experiment to measure the wavelength of monochromatic light;

2.4.10 use the equation $\lambda = \dfrac{ay}{d}$;

2.4.11 describe and explain diffraction phenomena at a single slit;

2.4.12 state qualitatively and draw diagrams to illustrate the effect of aperture size on diffraction; and

2.4.13 use the equation $d \sin \theta = n\lambda$ for a diffraction grating;

2.4.14 describe the use of a diffraction grating and a laser to measure wavelength;

Principle of Superposition

The Principle of Superposition states that the resultant displacement of the medium at any point in space, is the vector sum of the displacements that each wave would cause at that point at that time.

When the two waves overlap in phase they produce a wave of greater amplitude. This is known as **constructive interference**. The crests of each wave coincide exactly (as do the troughs) and so a wave with greater amplitude is produced.

When the crest of one wave coincides with the trough of the other wave the displacements of the two waves are in opposite directions. If the amplitudes are equal then they cancel each other. This is called total **destructive interference**.

If the amplitudes are not the same then when destructive interference takes place the resultant wave has a smaller amplitude.

Worked Example

Two sinusoidal transverse waves W_1 and W_2, each of the same amplitude, are incident simultaneously on a point P. Graphs A and B show how the displacement varies with time for waves W_1 and W_2 respectively.

(a) Draw the displacement–time graph for the resultant wave produced by the superposition of waves W_1 and W_2 and label it C.

(b) If the frequency of wave W_1 is 4×10^{15} Hz, what is the frequency of the resultant wave produced by the superposition of W_1 and W_2?

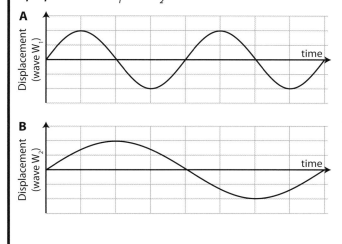

A

B

(a) To draw the resultant wave, the displacement of W_1 is added to the displacement of W_2 at each instant. The resultant is shown by the dots. When the dots are joined the shape of the resultant wave is obtained.

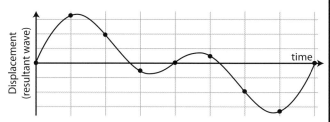

(b) The diagram below shows the waveforms for W_1, W_2 and their resultant W_{res}. W_{res} has a more complex shape, but this shape occupies the same time interval as that of W_2. So the frequency of the resultant wave W_{res} is the same as that of W_2, which is 2×10^{15} Hz, or half that of W_1.

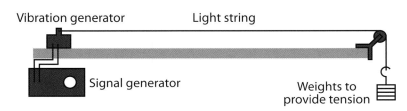

Standing Waves in Stretched Strings

These are a produced by the interference of **two waves**, of the **same type** and having the **same wavelength** but moving in **opposite directions**. Standing waves can be demonstrated using the apparatus opposite. The vibration generator vibrates up and down at a frequency which can be adjusted on the signal generator. The different **modes of vibration** which are observed are listed below.

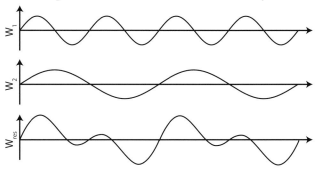

Vibration generator   Light string

Signal generator   Weights to provide tension

Fundamental

Frequency = f_1 and Wavelength $\lambda_1 = 2L$

$$v = f_1 \lambda_1 \qquad f_1 = \frac{v}{\lambda_1} = \frac{v}{2L}$$

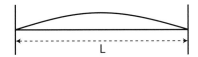

L

Next mode of vibration

Frequency = f_2 and Wavelength $\lambda_2 = L$

$$v = f_2 \lambda_2 \qquad f_2 = \frac{v}{\lambda_2} = \frac{v}{L} = 2f_1$$

Next mode of vibration

Frequency = f_3 and Wavelength $\lambda_3 = \frac{2L}{3}$

$$v = f_3 \lambda_3 \qquad f_3 = \frac{v}{\lambda_3} = \frac{3v}{2L} = 3f_1$$

Observe that the frequency, f_n, of the nth mode of vibration is $\frac{nv}{2L}$.

Nodes and Antinodes in Vibrating Strings

When a standing or stationary wave is created, some points along the wave are always at rest, their resultant displacement is always zero. These points are known as **nodes**. The distance between neighbouring nodes is always ½λ.

Between the nodes all the points are vibrating, but the amplitude of vibration varies. Midway between two nodal points the amplitude of vibration is a maximum. This point is called an **antinode**.

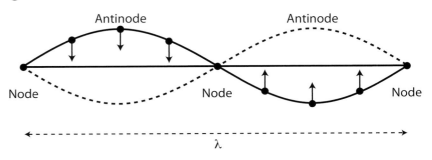

Standing Waves in Air Columns

Fixed length of air column – varying frequency

When the apparatus is set up as shown, sound waves from the speaker or tuning fork meet the reflected waves from the bottom of the air column and a standing wave is created. When a standing wave is produced the sound becomes much louder. Upon removing the glass tube there should be a very noticeable decrease in the loudness of the sound. There are different frequencies at which this happens – these are the different modes of vibration. The lowest frequency of sound that creates a standing wave for a particular length of air column is called the fundamental.

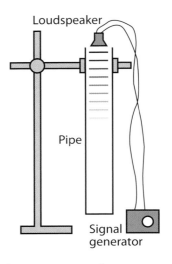

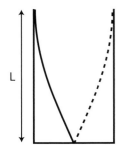
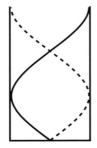
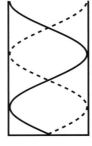
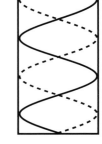

Fundamental

Frequency = f_1

and Wavelength $\lambda_1 = 4L$

$v = f_1\lambda_1 \quad f_1 = \dfrac{v}{4L}$

Next mode

Frequency = f_2

and Wavelength $\lambda_2 = \dfrac{4L}{3}$

$v = f_2\lambda_2 \quad f_2 = \dfrac{3v}{4L}$

Next mode

Frequency = f_3

and Wavelength $\lambda_3 = \dfrac{4L}{5}$

$v = f_3\lambda_3 \quad f_3 = \dfrac{5v}{4L}$

Next mode

Frequency = f_4

and Wavelength $\lambda_4 = \dfrac{4L}{7}$

$v = f_4\lambda_4 \quad f_4 = \dfrac{7v}{4L}$

Fixed frequency – varying length of air column

It is also possible to use a single tuning fork of fixed frequency and instead increase the length of the air column, using the apparatus below. If fundamental resonance occurs at length L, then the next highest mode of vibration occurs at length 3L, the next highest at 5L and so on.

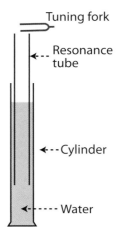

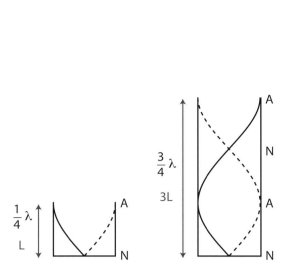

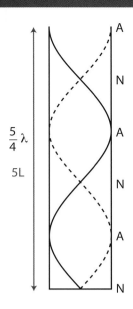

Measuring the Speed of Sound using a Resonance Tube

A vibrating tuning fork of known frequency is held over the open end of resonance tube. The inner glass tube is raised from minimum length until the sound becomes noticeably louder. The length of the air column is then measured. The procedure is repeated for several tuning forks of different known frequencies.

Since, at fundamental resonance, $v = f\lambda$ and $\lambda = 4L$, then $v = 4Lf$.

So, $L = \dfrac{v}{4f} = \dfrac{v}{4} \times \dfrac{1}{f}$

and hence the graph of L (y-axis) against $\dfrac{1}{f}$ (x-axis) is a straight line through the origin.

The gradient of this straight line is $\dfrac{v}{4}$. Thus to find the speed of sound it is only necessary to measure the gradient of the graph of L against $\dfrac{1}{f}$ and multiply it by 4.

The graph on the right shows the result of one such experiment. Measuring the gradient of this graph yields a figure of 86 m s⁻¹, and so the speed of sound obtained by this experiment was 344 m s⁻¹.

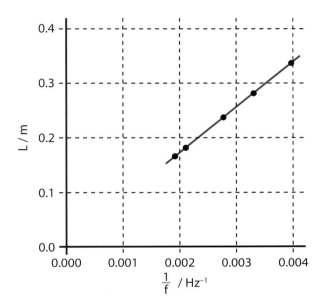

Coherence

To produce an interference pattern that is detectable, ie lasts long enough to seen or heard, the two sources of waves must be **coherent.** To be coherent the sources must produce waves of **the same wavelength or frequency** and have **a constant phase difference** between them. To make the difference between constructive and destructive more obvious, it is **best that the coherent sources are of equal amplitude.**

Interference of Sound Waves

In the diagram, S_1 and S_2 are two speakers. To achieve coherent sources of sound, the same signal generator powers each speaker, so that they produce sound waves of the same frequency and in phase. As the sound waves from each speaker spread out they cross. This creates places where the sound is loud (constructive interference) and between these there are places where the sound is soft (destructive interference).

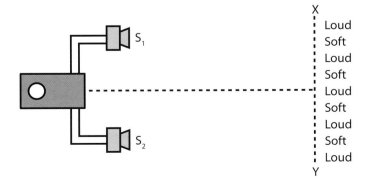

Interference of Light Waves - Young's Double Slit Experiment

Note that when using laser light, we do not darken the room. This would cause the pupils of student's eyes to dilate and increase the harm caused to the retina if laser light entered the eye. With all other light, we darken the room to see more clearly the faint interference fringes formed.

In this experiment, light from a laser is allowed to illuminate two narrow slits, as shown in the diagram. Laser light is coherent.

Each slit then acts as a coherent source of light waves. As the waves spread out interference is observed on a screen a distance (1 to 2 metres) away. The bright fringes are due to constructive interference and dark ones dues to destructive interference.

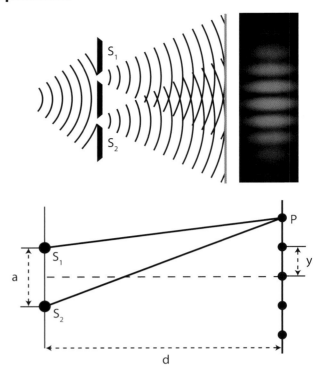

The second diagram shows the geometry of the two source interference experiment. S_1 and S_2 are two coherent sources of light, of wavelength λ, separated by a distance a. An interference pattern of alternate bright and dark fringes is seen on the screen. The separation of bright fringes is y.

The distance from the double slit to the screen is d. The point P is the location of a bright fringe. The waves reaching P from S_1 and S_2 have travelled different distances. If a whole number of wavelengths can fit into this path difference then constructive interference results since a crest from S_1 will arrive at the same time as a crest from S_2:

So for constructive interference:

Path difference $S_2P - S_1P = n\lambda$ where n = 0, 1, 2, 3...

However, if a whole number of wavelengths plus half a wavelength can fit into this path difference then destructive interference results since a crest from S_1 will arrive at the same time as a trough from S_2:

So for destructive interference:

Path difference $S_2P - S_1P = (n + \frac{1}{2})\lambda$ where n = 0, 1, 2, 3...

The wavelength can then be calculated using the formula:

$$\lambda = \frac{ay}{d}$$

where λ = wavelength in m
a = separation of the two slits in m
y = separation of bright fringes in m
d = distance to the screen in m

Worked Example

(a) What is monochromatic light?

(b) The diagram is a sketch of an arrangement used to measure the wavelength of light from a laser. Laser light is monochromatic.

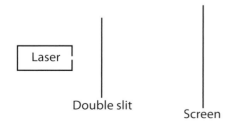

(i) Describe the pattern that will be seen on the screen.

(ii) The centres of the slits are 0.36 mm apart and the distance from the slits to the screen is 4.20 m. If the distance between adjacent intensity maxima is 7.4 mm, calculate the wavelength of the laser light.

(iii) State two ways in which the arrangement could be changed, using the same laser, so that the distance between the positions of maximum intensity seen on the screen would be increased.
(CCEA January 2011, amended)

(a) Monochromatic light is light of a single wavelength or frequency.

(b) (i) The pattern on the screen consists of a series of equally spaced alternate bright and dark bands or fringes.

   (ii) The wavelength can be calculated using $\lambda = \frac{ay}{d}$
   = $(0.36 \times 10^{-3} \times 7.4 \times 10^{-3}) \div 4.20$
   = 6.34×10^{-7} m
   Note that all distances are converted to metres. Thus wavelength in namometres is 6340 nm

   (iii) Re-arranging the equation gives $y = \frac{\lambda d}{a}$

   The distance y can be increased by increasing d, ie increasing the distance from the double slits to the screen. It can also be increased by reducing the value of a, ie by making the two slits closer together.

Diffraction

As waves go through a gap they **spread out**. This is called **diffraction**. Diffraction increases as the size of the gap is gradually decreased. The greatest diffraction happens when the size of the gap is **about the same as the wavelength of the wave**. The wavelength of the wave does not change as a result of diffraction.

Diffraction of sound takes place at an open door because the wavelength of sound is similar in size to the width of the door. Light has a much smaller wavelength so very narrow openings or slits are required to observe diffraction of light. A laser beam can be directed through a slit and onto a screen as shown below.

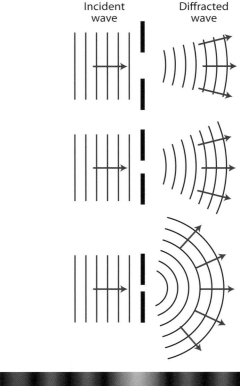

Incident wave   Diffracted wave

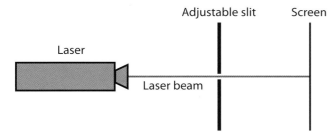

Laser

Adjustable slit   Screen

Laser beam

Diffraction of light waves at a single slit

Diffraction of the light waves at a single slit produces a diffraction pattern like that shown on the right.

Most of the energy of the light waves passes through in a narrow region giving rise to the bright central maximum. The width of the central maximum depends on the width of the slit and on the wavelength of the light used.

For the same slit width, blue light has a narrower diffraction pattern than red light because the wavelength of blue is less than the wavelength of red, as shown on the right.

The diagrams in the previous section showed how diffraction increases as the size of the gap is gradually decreased until is about the same size as the wavelength of the incident wave. **The greatest diffraction happens when the size of the gap is about the same as the wavelength of the wave.**

As the width of the slit decreases:

• the **width** of the diffraction pattern **increases** (so fewer maxima are observed)

• the **heights** of the maxima **decrease** (because less energy arrives on the screen)

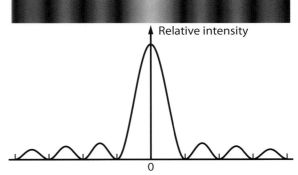

Relative intensity

0

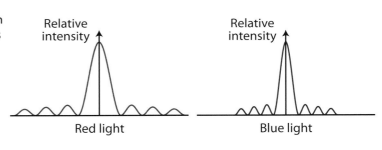

Relative intensity

Relative intensity

Red light   Blue light

Using a Diffraction Grating to Measure the Wavelength of Light

A transmission diffraction grating consists of a very large number of equally spaced parallel lines scratched on a glass. When monochromatic light, such as that from a laser, passes through the grating and on to the screen, a series of lines is observed. Each line corresponds to a different "order", n.

When all the waves spreading out from all the slits are added up, they cancel out everywhere except in certain directions along which all the crests of all the waves exactly coincide and add up constructively, creating the observed lines. These particular directions are determined by the wavelength of the light, λ, and the distance d, between centres of adjacent slits in the grating, known as the grating spacing. The relationship between the angle of diffraction θ and the distance between adjacent slits, d, is $d \sin\theta = n\lambda$.

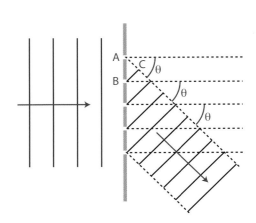

Measuring the wavelength therefore involves the practical measurement of the angles, θ_n, for each order. With reference to the diagram on the right, we see that $\theta_n = \tan^{-1}(y_n/D)$ where y_n is the distance on the screen between the line of order n and the axis of symmetry. Typically $D \approx 2$ m and is measured with a metre stick, while $y \approx$ several mm and is measured with a 30 cm ruler. The wavelength of the light being used is therefore the gradient of the graph of $d \times \sin\theta$ (y-axis) against n (x-axis).

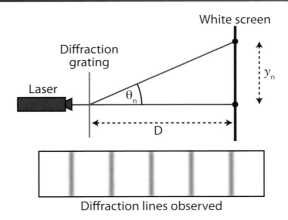

Diffraction lines observed

Exercise 16

1. In a normal six string guitar the top string is tuned so that its lowest natural frequency is 82 Hz when the full length of the string vibrates. The diagram opposite represents the guitar but only the top string has been shown.

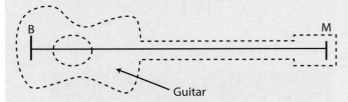

(a)(i) Copy the diagram and draw the first mode of vibration (fundamental) for the string.

(ii) Label every node with an N and every antinode with an A.

(b)(i) If the distance between B and M is 0.84 m, what is the wavelength of the first mode of vibration of the standing wave on the string?

(ii) To produce a note of higher frequency the guitarist places one finger at a point X on the string. The string cannot move at that point and the vibrating length is effectively reduced. He then plucks the string with another finger between X and B. The note obtained has a fundamental frequency of 328 Hz. Calculate the distance X to B.

(c) Guitarists are able to produce different modes of vibration on the same length of string by lightly touching the string. This creates a node at the point touched but does not reduce the effective length of the string that is vibrating.

Copy the diagram below and sketch the simplest mode of vibration that results when a guitarist touches the string at position F. The distance FM is 0.28 m and the distance BM is 0.84 m.

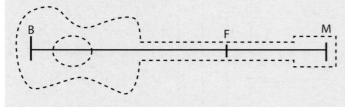

(CCEA June 2011)

2. Residents of a housing development near a busy motorway are shielded from the noise by a barrier. The diagram below is a plan view of the situation showing houses (A to J), sound wavefronts from the motorway and the barrier.

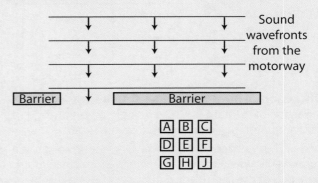

(a) Draw a copy of the diagram, and continue the path of the sound wavefront, as it passes the barrier, to show it in its next three positions.

(b) State and explain what will happen to the shadow zone (the region behind the noise barrier into which no sound enters) when the mean wavelength of the sound from the motorway increases.

(CCEA June 2011)

3. Laser light is incident upon an opaque slide on which two there are two transparent slits, S and T. The light transmitted results in an interference pattern on the screen.

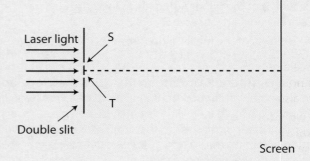

(a) The light at S and T is coherent. What does this mean?

(b) On the interference pattern seen on the screen there is a series of bright and dark bands. Explain how these bands arise.

(c) The distance between the centres of the first and the seventh dark band is 24.6 mm and the separation of the double slits is 0.60 mm. The distance between the double slit and the screen is 3.9 m.

    (i) Calculate the fringe width, y.

    (ii) Calculate the wavelength of the laser light.

    (iii) What colour is the laser light?

4. In a Young's double slit experiment using violet light of wavelength 400 nm the separation of adjacent fringes is 500 µm. The violet lamp is replaced by one emitting light red light of wavelength 600 nm. The remainder of the apparatus is unchanged. By how much is the separation of adjacent fringes changed as a result?

5. When observing a Young's interference pattern obtained using monochromatic light from a sodium lamp, it is a good idea to pull the window blinds to darken the room.

When observing a Young's interference pattern obtained using laser light, it is a not a good idea to darken the room completely. Explain why this is so.

6. Monochromatic light is shone normally through a transmission diffraction grating with 250 lines per millimetre. The angle between the rays in the second order is 36.8°. Calculate

(a) the distance, d, between adjacent slits;

(b) the angle of diffraction, θ, in the second order;

(c) the wavelength of the monochromatic light used;

(d) the highest order of diffraction which can be observed when this grating is used with this light source.

2.5 Quantum Physics

Students should be able to:

2.5.1 recall and use the equation $E_{photon} = hf$;

2.5.2 use the photon model to explain the photoelectric effect qualitatively using the terms photon energy and work function;

2.5.3 use the equation $\frac{1}{2}mv_{max}^2 = hf - hf_0$;

2.5.4 demonstrate an understanding that electrons exist in energy levels in atoms;

2.5.5 recall and use the equation $hf = \Delta E$;

2.5.6 provide a simple explanation of laser action, using the terms population inversion and metastable state;

2.5.7 demonstrate an understanding of the production of X-rays by the process of electron movement between energy levels;

2.5.8 describe the physical principles of CT scanning and conventional X-rays;

Photoelectric Emission

Electrons are ejected from the surface of a metal when electromagnetic radiation of sufficiently high frequency falls on it. This is called the **photoelectric effect**. The electrons emitted by this process are called **photoelectrons**.

Wave theory cannot explain some crucial experimental observations associated with this emission:

1. The effect is instantaneous, that is there is no time delay between the radiation falling on the surface and the onset of photoelectric emission. Wave theory predicts a time delay of around 90 minutes.

2. There is a **threshold frequency** for each metal below which emission will not take place, regardless of the light's intensity.

To explain this effect we need to use the **photon model** of light. In this model we regard light as packets or **quanta**, each possessing a discrete quantity of energy. These quanta are called **photons**. The energy of a photon is given by:

E = hf or $\frac{hc}{\lambda}$ where E = energy of the photon in J

                      h = Planck's constant = 6.63×10^{-34} Js

                      f = frequency of the incident radiation in Hz

                      c = speed of light = 3×10^8 m s^{-1}

                      λ = wavelength of the radiation in m

It is common to give the energy of a photon in **electron-volts**. An electron-volt is the energy gained by an electron when it is accelerated through a potential difference of 1 volt. 1eV = 1.6×10^{-19} J. To explain the photoelectric effect Einstein assumed that not only were light and other forms of electromagnetic radiation emitted in whole numbers of photons, but that they were also absorbed as photons.

The **work function** Φ is defined as **the minimum quantity of energy needed to liberate electrons from the surface of a metal and to just allow it to escape to an infinite distance from the metal**. Einstein proposed that a photon of energy will cause the emission of an electron from the metal if the energy of the photon is equal to or greater than the work function of the metal.

If the photon's energy is greater than the work function of the metal then the difference appears as kinetic energy of the ejected electron. Since the work function is the minimum energy needed to eject an electron from the metal this means that the electrons that are ejected will have a range of kinetic energy from zero to a maximum.

Einstein's **Photoelectric Equation** can be written as:

Incident photon energy – Work function energy = Maximum kinetic energy of the electrons

or **hf – Φ = $\frac{1}{2}mv_{max}^2$**

The frequency of the electromagnetic radiation that just liberates an electrons is called the threshold frequency f_0 and is equal to the work function Φ of the metal. So, $hf_0 = \Phi$. This allows us to write Einstein's Photoelectric Equation in a more useful form:

$$hf - hf_0 = \tfrac{1}{2}mv_{max}^2$$ where   f = frequency of the incident radiation in Hz

f_0 = threshold frequency for the given metal in Hz

$\tfrac{1}{2}mv_{max}^2$ = maximum kinetic energy of the emitted photoelectrons in J

Worked Example

Sodium has a work function of 2.28 eV.

(a) Calculate the minimum frequency of an incident photon which will just cause photoemission in sodium.

(b) Find the maximum velocity of the photoelectrons emitted when freshly cut sodium metal is illuminated with monochromatic light of wavelength 500 nm.

(c) Explain why many of the photoelectrons have a velocity less than this maximum.

(d) The intensity of the incident radiation is now doubled and the wavelength is increased to 600 nm. What effect does this have on the number of photoelectrons emitted peer second from the surface of the metal?

(a) $f_0 = \Phi \div h = (2.28 \times 1.6 \times 10^{-19}) \div 6.63 \times 10^{-34} = 5.53 \times 10^{14}$ Hz

(b) $hf = \Phi + \tfrac{1}{2}mv^2$, so $h(c \div \lambda) = \Phi + \tfrac{1}{2}mv^2$

$6.63 \times 10^{-34} \times (3 \times 10^8 \div 500 \times 10^{-9}) = (2.28 \times 1.6 \times 10^{-19}) + \tfrac{1}{2} mv^2$

So: $\tfrac{1}{2}mv^2 = 3.978 \times 10^{-19} - 3.648 \times 10^{-19} = 3.3 \times 10^{-20}$ J

Hence $v^2 = 6.6 \times 10^{-20} \div 9.11 \times 10^{-31}$

(since $m = 9.11 \times 10^{-31}$, ie the mass of an electron)

giving $v = 2.69 \times 10^5$ m s^{-1}.

(c) Many electrons are liberated from atoms which are several atomic diameters beneath the surface. As they travel towards the surface these electrons lose kinetic energy in collisions with the atoms in the metal lattice. Only those electrons liberated from atoms on the surface will have maximum kinetic energy.

(d) Since $f_0 = 5.53 \times 10^{14}$ Hz, the maximum wavelength which can achieve photoelectric emission, λ_0 is given by: $\lambda_0 = 3 \times 10^8 \div 5.53 \times 10^{14} = 542$ nm, so no emission occurs when the surface is illuminated with 600 nm light, regardless of its intensity.

The Emission Spectrum

When sunlight is made to pass through a triangular glass prism, as shown in the diagram, a spectrum is obtained. We call such a spectrum continuous. We can also obtain a continuous spectrum from hot filament lamps.

However, when we look at the light from a gas discharge lamp containing a gaseous element such as sodium vapour or neon, we obtain a very limited range of wavelengths (colours) as shown in Figure 2.

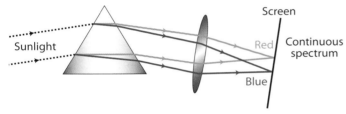

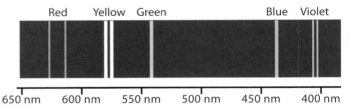

This is called a **line emission spectrum** and consists of a series of discrete wavelengths. To explain this type of emission spectrum we have to apply the idea of quantisation to the electron in orbit around the nucleus of the atoms.

An electron has a fixed amount of energy in each orbit, those being closest to the nucleus having the least energy and those most distant from the nucleus having the most energy.

An electron can move from one energy level to a higher energy level by absorbing a photon of energy equal to the energy difference between the two states. This process is called **excitation**.

An electron in an excited state can move to a lower energy level, by emitting a photon of light of energy exactly equal to the energy difference between the two states would be emitted. This process is called **relaxation**.

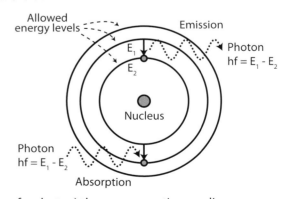

For both excitation (absorption of a photon) and relaxation (emission of a photon) the same equation applies:

$$\Delta E = hf \text{ or } \frac{hc}{\lambda}$$ where ΔE = energy difference between the two levels in J

h = Planck's constant = 6.63×10^{-34} Js

f = frequency of the radiation in Hz

c = speed of light = 3×10^8 m s^{-1}

λ = wavelength of the radiation in m.

You should appreciate that the equation above is an example of the principle of conservation of energy in a form that applies to electron transitions between orbits.

Energy Level Diagrams

The diagram on the right is an **energy level diagram** showing the main electron transitions in hydrogen. The lowest energy level or ground state has value of –13.6 eV. Higher energy levels are less negative, –3.4 eV, –1.5 eV and so on. The energy levels are often given in electron–volts where 1 eV = $1.6×10^{-19}$ J.

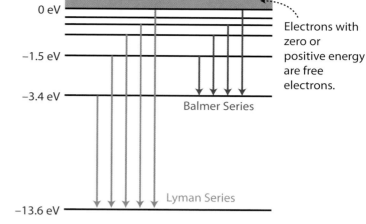

The **energy levels are negative** since energy has to be supplied to remove the electron completely from the atom. A stationary free electron after ionization is said to have zero energy.

The **longest wavelength** in the Lyman series corresponds to a photon with minimum energy. This photon is emitted when the electron moves from level with energy –3.4 eV to the ground state level with energy –13.6 eV.

The energy of the emitted photon = 13.6 – 3.4 = 10.2 eV = $1.63×10^{-18}$ J.

The frequency f = ΔE ÷ h giving a value of $2.46×10^{15}$ Hz and a wavelength of 121.9 nm.

The **shortest wavelength** in this series corresponds to a photon with maximum energy. This photon is emitted when the electron moves from an energy of 0 to the ground state level of energy –13.6 eV.

The energy of the emitted photon = 13.6 – 0 = 13.6 eV = $2.18×10^{-18}$ J.

The frequency f = ΔE ÷ h giving a value of $3.29×10^{15}$ Hz and a wavelength of 91.2 nm.

The reader should verify that all transition to the –13.6 eV energy level give rise to radiation in the ultraviolet part of the electromagnetic spectrum, while all transitions in the Balmer series are in the visible part of the spectrum.

Worked Example

(a) What is the evidence that the electron in the hydrogen atom can only exist in states of discrete energy?

(b) Explain what is meant by the "ground state".

The lowest energy states in hydrogen are at –13.6 eV and –3.4 eV.

(c) Calculate the wavelength of the incident radiation which will cause an electron in the lowest energy state to become excited.

(d) Explain why hydrogen does not absorb visible light when its electron is in the ground state.

(a) The emission line spectrum from a hydrogen discharge lamp has very limited range of discrete wavelengths. This is because electrons are confined to circular orbits of discrete energy. An electron in an excited state can relax by moving to a lower energy level, by emitting a photon of light of energy exactly equal to the energy difference between the two states would be emitted.

(b) The ground state is the lowest energy level that the orbiting electron can have and corresponds to an orbit closest to the nucleus.

(c) E = E_1 – E_2 = $(-3.4) - (-13.6) × 1.6×10^{-19}$ J = $1.63×10^{-18}$ J
λ = hc ÷ E = $(6.63×10^{-34} × 3×10^8) ÷ 1.63×10^{-18}$ = 122 nm

(d) The energy of a photon of visible light does not equal the energy level difference between the ground state and the next excited energy level so it cannot be absorbed.

Laser Action

In nature, there are normally many more electrons in the ground state than in any excited state. The length of time an electron spends in an excited state is typically 10^{-8} s. It then makes a transition to a lower energy level and a photon of light is emitted. This is a random process and is called **spontaneous emission**.

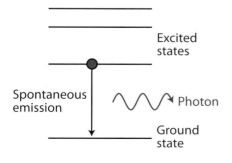

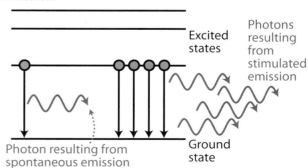

In a laser, we need to have more electrons in the excited state than in the ground state. This is called a **population inversion**. Electrons can exist in some states for a much greater time than in normal states. These are called **metastable states** and the time spent can be around 10^{-3} s. This is **100 000 times longer than normal**. This gives us time to have more electrons in this excited state than in the ground state. However electrons in a metastable excited state can be induced to make a transition by a photon of energy equal to the difference between the levels passing close to the atom. This is known as **stimulated emission**. This 'inducing' photon results from spontaneous emission. Stimulated emission is extremely unlikely to occur in nature.

A diagrammatic view of laser action

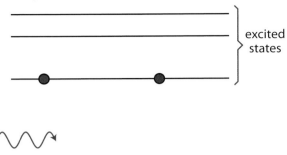

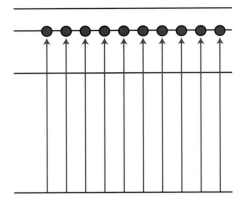

Stimulated emission is very unlikely because there are many more electrons in the ground state than in the excited state. Excitation is therefore more probable than stimulated emission.

The first stage in achieving a population inversion is to cause the majority of the electrons in the ground state to move to an excited state by optical pumping. But the excited state is short lived.

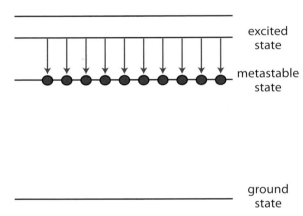

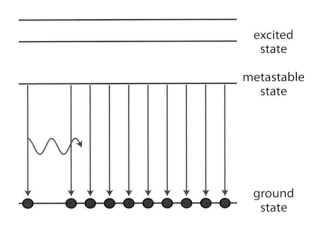

Electrons in the excited, short lived state quickly relax to the metastable state, where they can remain for much longer than in an ordinary excited state.

The photon generated by the first (and spontaneous) relaxation from the metastable state causes the other electrons in the metastable state to return to the ground state by stimulated emission.

The photons produced by stimulated emission are not only coherent, but they are exactly in phase with one another. So the wave produced is greatly amplified, hence the acronym 'laser' (light amplification by stimulated emission of radiation).

X-Ray Production

X-rays are produced by high speed electrons striking metal targets. The electrons are emitted by a **cathode** which is heated to a very high temperature (white heat). This heating is produced by an electric current passing through the cathode. The electrons are then accelerated toward a metal **anode** using a very high voltage (typically 100 kV). The space inside the X-ray tube is a vacuum.

At the centre of the anode is a metal target. This target is made of metal of a high melting point and high atomic number. Around **0.5%** of the electrons produce **X-rays**, the other 99.5% simply heat the anode. The anode therefore needs to be constantly cooled. In the case of the rotating anode tube shown, this is achieved by conduction.

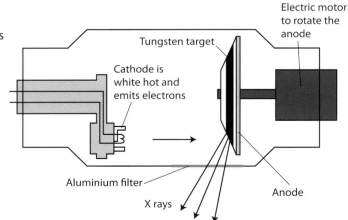

Compared to a fixed anode, the rotating anode presents a larger area for the electrons to strike. This larger area allows for a more efficient removal of heat by conduction. The anode also radiates heat which is absorbed by the surrounding glass envelope which then transfers heat to the surroundings by radiation and convection.

Because the elements used as the target have high atomic numbers, the electron shells are generally filled with the total complement of electrons. An incoming electron will knock an electron out of these filled electron shells, as shown in the diagram on the right. The vacancy left is immediately filled by an electron from a higher energy shell dropping down to a lower energy shell. It loses its energy as an X-ray of very specific energy or wavelength.

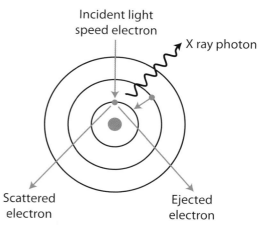

In all atoms of low atomic number transitions to the innermost shell give rise to the emission of ultraviolet light. It is for this reason that metals like aluminium are quite unsuitable as targets in an X-ray tube. Their emission lines (emission spectra) have wavelengths higher than X-rays.

Conventional X-rays

Within the human body, denser objects, such as bones, absorb more radiation than less dense material, such as muscle. Traditionally the detector used was very sensitive photographic film, but modern hospitals now use digital detectors, which are faster, cheaper, use less non-renewable materials and produce no waste.

The subject to be X-rayed is placed between the X-ray source and the detector. The radiographers withdraw to a place of safety behind a lead screen and the X-rays are generated. Bones appear white on X-ray images. This is because the bones absorb more X-rays and their 'shadow' is cast on the X-ray film.

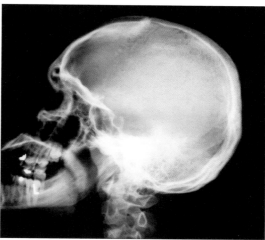

Computed Tomography or CT Scanning

Computed tomography (CT) imaging, also known as 'CAT scanning' (computed axial tomography), was developed in the early to mid 1970s. CT has the unique ability to image a combination of soft tissue, bone and blood vessels. Today CT enables the diagnosis of a wide range of illness and combines the use of a digital computer with a rotating X-ray device to create detailed cross-sectional images or 'slices' of the different organs and body parts, such as the lungs, liver, kidneys, brain, spine and blood vessels.

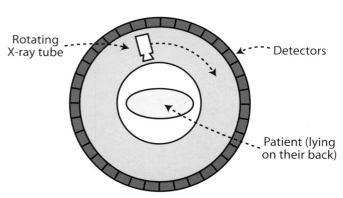

CT images allow us to see soft-tissue structures like the valves of the heart or grey matter in the brain. CT is an invaluable tool in cancer diagnosis and is often the preferred method for diagnosing lung, liver and pancreatic cancer. CT imaging has a role in the detection, diagnosis and treatment of heart disease, acute stroke and vascular diseases which can lead to stroke. Additionally, CT can be used to measure bone mineral density for the detection of osteoporosis.

Disadvantages of CT scanning

- Computed tomography is considerably more expensive than conventional radiography.
- The X-ray dose from CT is much larger than for conventional radiography.
- For scans of the abdomen patients often have to drink several large cupfuls of contrast agent about an hour before scanning. The contrast agent has a high density so in the scan the entire bowel is highlighted in white and readily distinguished from possible intra-abdominal masses.
- Because of the large X-ray dose from CT other methods are always considered before CT is carried out on children, as they are particularly sensitive to ionising radiation.

Exercise 17

1. A metal with a photoelectric work function of 3.7×10^{-19} J is illuminated with light of wavelength 476 nm.

   (a) Calculate the maximum kinetic energy of the emitted photoelectrons. Give your answer in eV.

   Loss of electrons by photoelectric emission causes the surface to acquire a positive potential.

   (b) What will this potential if it just prevents the further photoelectric emission when illuminated with light of wavelength 476 nm?

2. Light of wavelength 450 nm illuminates a calcium surface which a has a work function of 4.3×10^{-19} J.

   (a) Show that photoelectric emission will occur.

   (b) In what way, if at all, will the number of electrons emitted per second from the surface change if the wavelength reduces to 400 nm and the intensity of the incident radiation remains the same as before?

3. (a) The red line in the emission spectrum of hydrogen has a wavelength of 655 nm. Use the energy level diagram below to identify the transition which brings about this emission. Copy the diagram and indicate with an arrow the direction of the transition.

   ```
     0 eV · · · · · · · · · · · · · · · · · ·
    −0.85 eV ————————————————————
    −1.51 eV ————————————————————

    −3.40 eV ————————————————————



    −13.6 eV ————————————————————
   ```

   (b) Use the diagram to show that all relaxations to the energy level −1.51 eV in hydrogen result in the emission of infra-red light.

4. (a) State what happens when an electron moves from one energy level to another of lower energy.

   (b) What is the difference between spontaneous and stimulated emission of light?

   (c) Explain why stimulated emission does not occur naturally.

   (d) Give a brief description of what is required to bring about stimulated emission.

   (e) State three differences between laser light and sunlight.

5. (a) Draw a labelled diagram to show the main parts of a modern X-ray tube.

   (b) Most of the kinetic energy of the electrons striking the target in an X-ray is converted into heat. Discuss briefly two ways by which this heat is removed from the target.

   (c) X-ray spectra typically show discrete lines at particular wavelengths. Give a brief account of the source of these discrete lines.

   (d) A particular X-ray tube uses a tungsten target and the electrons strike the target with a kinetic energy of 100 keV.

   (i) Use the Principle of Conservation of Energy to show that the minimum wavelength of the X-rays from this tube is 1.24×10^{-11} m.

   (ii) What change in the X-ray spectrum would be observed if copper was used instead of tungsten as the target?

6. (a) A patient is to undergo a CT scan. Describe what happens.

   (b) Give three reasons why a doctor might decide against giving a patient a CT scan.

2.6 Wave–Particle Duality

Students should be able to:

2.6.1 categorise electromagnetic wave phenomena as being explained by the wave model, the photon model or both;

2.6.2 describe electron diffraction; and

2.6.3 use the de Broglie equation $\lambda = \dfrac{h}{p}$.

To explain some aspects of light behaviour, such as interference and diffraction, it is treated as a wave. To explain other aspects it is treated as being made up of particles (photons). Light exhibits **wave–particle duality**, because it exhibits properties of **both** waves **and** particles:

Phenomenon	Can be explained in terms of waves	Can be explained in terms of photons
Reflection of light	Yes	Yes
Refraction of light	Yes	Yes*
Interference of light	Yes	No
Diffraction of light	Yes	No
Polarisation of light	Yes	No
Photoelectric effect	No	Yes

*Refraction can be explained using the particle nature of light, but not at a simple level. It requires some very advanced physics known as Quantum Electrodynamics.

Electron Diffraction

In 1924, Louis de Broglie (pronounced de Broy), suggested that matter might also have a dual nature. He suggested that a moving particle has an associated wavelength given by the equation:

$$\lambda = \frac{h}{p}$$

where λ = de Broglie wavelength, in m
  h = Planck's constant, 6.63×10^{-34} Js
  p = momentum of the particle, in kg m s^{-1}

Calculations show that if electrons were accelerated through 100 V then their momentum would indicate a wavelength of around 10^{-10} m. Using X-ray diffraction it had been shown that the distance between atoms in crystals is around 10^{-10} m. So de Broglie thought it might be possible to use the layers of atoms in a crystal to produce diffraction and interference effects using 100 V electrons also.

Electron diffraction confirming de Broglie's ideas was first carried out in 1927 by Clinton Davisson and Lester Germer in the USA. It can be demonstrated in schools using the apparatus shown. A crystal of graphite is used to diffract a beam of electrons. The pattern is observed on a fluorescent screen.

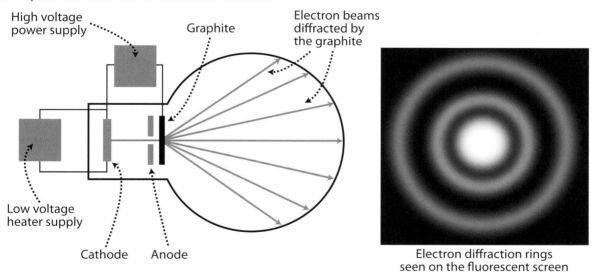

Electron diffraction rings seen on the fluorescent screen

Observe that the electron diffraction pattern is almost identical to that obtained by X-ray diffraction. Where lots of electrons reach the fluorescent screen a bright ring is seen. This is a region of constructive interference. In other areas the absence of light indicates a region of destructive interference where few electrons reach the screen.

If the accelerating voltage is increased, the speed of the diffracting electrons (and hence their momentum) also increases. The rings then **become narrower and have a smaller radius**, showing that the **wavelengths of the electron waves decrease with increasing momentum**.

Electrons also exhibit the same kind of interference pattern as light does when they pass through a double slit (Young's experiment) of the correct dimensions.

Why then do we not observe other moving objects, such as cars, trains and buses, displaying wave-like properties? The answer is that their mass (and hence their momentum) is so large and Planck's constant is so small, that the **wavelength of these objects is much too small to produce observable interference and diffraction effects**.

Exercise 18

1. (a) A certain electron in a hydrogen atom has a kinetic energy of 2.2×10^{-18} J.
    (i) Show that this electron has a speed of 2.2×10^6 m s^{-1}.
    (ii) Calculate the magnitude of the momentum of this electron.
    (iii) Use de Broglie's equation to find the wavelength of this electron.
   (b) Explain why a bullet of mass 25 g travelling at 400 m s^{-1} through the air is unlikely to produce observable diffraction effects.

2. A particle moving at 2×10^6 m s^{-1} has a de Broglie wavelength of 3.64×10^{-10} m. Find the probable identity of the particle.

3. A beam of protons and a beam of electrons are both accelerated to a speed of 2 Mm s^{-1}.
   (a) Which particles have the longer de Broglie wavelength?
   (b) Calculate the ratio of the wavelength of the electrons to that of the protons.
   Now suppose a beam of protons and a beam of electrons are both accelerated until the kinetic energy of each is the same, say 10 keV.
   (c) Calculate the ratio of the wavelength of the electrons to that of the protons when each has the same kinetic energy.

4. The wavelength of moving electrons depends on their speed. Copy the graph axes shown and sketch the graph which shows the nature of this relationship.

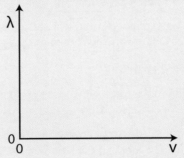

5. Calculate the de Broglie wavelength of electrons accelerated through a potential of 100 V. (Hint: Such electrons have a kinetic energy of 100 eV.)

2.7 Astronomy

Students should be able to:

2.7.1 recall, demonstrate an understanding of and apply the classical equations for Doppler shift to find the wavelength of the waves received by a stationary observer from a moving source;

2.7.2 demonstrate an understanding of the difference between red shift and Doppler red shift;

2.7.3 calculate the red shift parameter, z, of a receding galaxy using the equation $z = \dfrac{\Delta\lambda}{\lambda}$ and use the equation $z = \dfrac{v}{c}$ to find the recession speed v, where $v \ll c$;

2.7.4 use Hubble's Law $v = H_0 d$ to estimate the distance, d, to a distant galaxy, given the value of its speed of recession, v, and the Hubble constant, $H_0 \approx 2.4 \times 10^{-18}$ s^{-1}; and

2.7.5 recall and use $T = \dfrac{1}{H_0}$ to estimate the age of the universe.

Classical Doppler Effect

If a wave source is moving, the crests of its waves get bunched together in front of the wave source. If the wave crests are bunched together, their wavelength decreases. On the other side of the source, the waves spread out and the wavelength increases. This is known as the **Doppler Effect**.

The Doppler Effect explains why the sound of a siren from a fire engine appears to have a bigger pitch (smaller wavelength) as it approaches us and a smaller pitch (bigger wavelength) as it moves away from us.

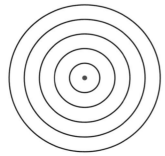

Wavefronts for a stationary source.

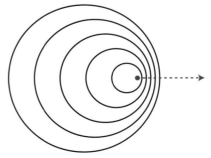

Wavefronts for a source moving to the right.

Classical Doppler Effect Equations

In classical physics, where the speed of source relative to the medium is lower than the speed of waves in the medium, the relationship between observed frequency f and emitted frequency f_0 is given by:

$$f = \left(\frac{v_w}{v_w + v_s}\right) f_o$$

where v_w is the velocity of waves in the medium, and
v_s is the velocity of the source relative to the medium; positive if the source is moving away from the receiver (and negative in the other direction).

Taking the reciprocal of both sides and then multiplying by v_w gives:

$$\lambda = \left(1 + \frac{v_s}{v_w}\right) \lambda_o$$

where λ is the observed wavelength, and
λ_o is the emitted wavelength

These two equations should be memorised.

Note that in the AS course we are only concerned with Doppler Effect **arising from movement of the source**. We are never concerned with Doppler Effect due to movement of the receiver.

Worked Example

A car travelling at a speed of 30 m s^{-1} sounds its horn which emits a note of frequency 100 Hz. If the speed of sound in air is 330 m s^{-1}, calculate the apparent frequency and wavelength of the note as heard by a person some distance in front of the car.

Since the source is moving towards the receiver, we take v_s as -30 m s^{-1}.

Using $f = \left(\frac{v_w}{v_w + v_s}\right) f_o = \left(\frac{330}{330 - 30}\right) \times 100 = 110$ Hz

The emitted note has wavelength $\lambda_o = 330 \div 100 = 3.3$ m.

Again we take v_s as -30 m s^{-1} because the source is moving towards the receiver.

Using $\lambda = \left(1 + \frac{v_s}{v_w}\right) \lambda_o = \left(1 - \frac{30}{330}\right) \times 3.3 = 3.0$ m

---

Cosmological Doppler Effect in Astronomy

Edwin Hubble observed that wavelengths of the light emitted by hydrogen atoms in distant galaxies were all longer than those observed for hydrogen in the laboratory. This is what we call **red shift**. Hubble concluded that this red shift must be because the stars in these galaxies were moving away from us at great speed. The universe was expanding.

Cosmologists have a different perspective. They think of the expanding universe as **the expansion of space itself**. So light waves from a distant galaxy are being stretched as the space between us and that galaxy increases. This is called **cosmological red shift** to distinguish it from red shift produced by sources that are moving through space.

Physicists define a quantity called the **cosmological red shift parameter, z**, by the equations:

$$z = \frac{\Delta\lambda}{\lambda} = \frac{\Delta f}{f} = \frac{v}{c}$$

where λ is the emitted wavelength, in m
f is the emitted frequency, in Hz
$\Delta\lambda$ is the difference between the observed and the emitted wavelengths
Δf is the difference between the observed and the emitted frequencies
v is the velocity of the source in the observer's direction, in m s^{-1}
c is the speed of the light waves in the medium, in m s^{-1}

The equation is only valid if v << c. (If v approaches the speed of light, then the equation must be modified in accordance with Einstein's theory of Special Relativity).

Worked Example

There is an absorption line in the spectrum of calcium which has a wavelength 393.3 nm. The same line in the spectrum of the light from a distant galaxy has a wavelength of 394.1 nm.

(a) Calculate the z-parameter for this galaxy.

(b) Calculate the speed with which this galaxy is moving with respect to the Earth.

(c) In what direction is the galaxy moving relative to the Earth?

(a) $z = \frac{\Delta\lambda}{\lambda} = \frac{(394.1 - 393.3)}{393.3} = 0.002$ (note that z has no unit)

(b) $v = zc = 0.002 \times 3\times10^8 = 6\times10^5$ m s^{-1}

(c) Since there is an increase in the observed wavelength (red shift), the galaxy is moving away from the Earth. We say the galaxy is receding with a recession speed of 6×10^5 m s^{-1}.

Hubble's Law

Astronomers have observed that apart from a few very close ones, all galaxies show a red shift. This means that they are moving away from us. It has been observer that a **graph of recessional velocity, v against distance, d is a straight line through the origin indicating direct proportion**. This graph gives rise to Hubble's Law, named in honour of Edwin Hubble who did much to confirm the relationship. Expressed as an equation, Hubble's Law is written:

$v = H_o d$ where   v is the recessional velocity, in m s^{-2}
d is the distance between Earth and the galaxy in question, in m
H_o is the Hubble parameter (also called the Hubble constant)

Many physicists prefer to call H_o the Hubble parameter, rather than the Hubble constant, as there is evidence to believe that the rate at which the universe is expanding, and hence the value of H, has changed over time. However, note that the CCEA specification uses the term Hubble constant. In SI units the approximate value of H_o is 2.4×10^{-18} s^{-1}.

Worked Example

The H-alpha line in the visible spectrum of hydrogen is observed in the laboratory to have a wavelength of 656.0 nm. The same line in the light from the Andromeda Galaxy has a wavelength of 655.8 nm. Calculate

(a) the speed and direction of Andromeda relative to our galaxy.

(b) how far away Andromeda is from the Milky Way.

(a) Note that the direction of Andromeda's motion is towards the Milky Way because its light shows a **blue shift**. This is quite unusual for galaxies.

Using $z = \dfrac{\Delta\lambda}{\lambda} = \dfrac{(655.8 - 656.0)}{656.0} = -3.05 \times 10^{-4}$

The minus sign tells us the direction of Andromeda's motion.

$v = zc = 3.05 \times 10^{-4} \times 3 \times 10^{8} = 9.15 \times 10^{4}$ m s^{-1}

(b) $d = \dfrac{v}{H_o} = 9.15 \times 10^{4} \div 2.4 \times 10^{-18} = 3.81 \times 10^{22}$ m

---

The Age of the Universe

Almost all physicists today accept the Big Bang Theory as the best explanation as to the origin of the universe. The idea is that the Universe began from a single point, a singularity. Now, if you imagine time in the Universe running backwards, how long would it take a distant galaxy to reach you? Answering this question tells you how long ago it is since all the galaxies were together in the same place, ie how long ago the Big Bang occurred.

Time, T, taken for a galaxy travelling at speed v to travel a distance d is:

$T = \dfrac{d}{v} = \dfrac{1}{H_o}$ (from Hubble's Law).

Therefore, the equation $T = \dfrac{1}{H_o}$ gives us an estimate of the age of the Universe.

So, assuming that the value of the H_o parameter is constant, $T = 1 \div 2.4 \times 10^{-18}$ s $= 4.17 \times 10^{17}$ s $= 13.2$ billion years.

Exercise 19

1. A loudspeaker, which emits a note of frequency 250 Hz, is whirled in a vertical circle of radius at a constant speed of 2.00 m s^{-1}. Use the classical Doppler equations to calculate the maximum and minimum frequencies detected by a stationary observer if the speed of sound is 330 m s^{-1}. Give your answers to 3 significant figures.

2. On a day when the speed of sound in air is 330 m s^{-1}, a car sounds its horn. The horn emits sound of frequency 100 Hz. A stationary observer, some distance behind the car, detects a sound of frequency 88 Hz. Calculate the speed of the car.

3. There is an absorption line in the ultraviolet spectrum of hydrogen which has a wavelength 122.2 nm. The same line in the spectrum of the light from a distant galaxy has a wavelength of 124.2 nm.

(a) Calculate the z-parameter for this galaxy.

(b) Calculate the speed with which this galaxy is moving with respect to the Earth.

(c) In what direction is the galaxy moving relative to the Earth?

4. The H-alpha line in the visible spectrum of hydrogen is observed in the laboratory to have a wavelength of 656.0 nm. The same line in the light from a distant galaxy has a wavelength of 657.8 nm. Calculate how far away this galaxy is from the Milky Way.

Answers

Note: When you have to perform calculations on a set of measurements, the result should be given to the same number of significant figures (sf) as the initial values.

Exercise 1 – *Unit 1.1*

1. (a) Use the equation $F = ma$. m = mass, measured in kg . a = acceleration, measured in m s^{-2}. The SI base units are therefore kg m s^{-2}.

   (b) $P = F \div A$. From (a) the SI units of force are kg m s^{-2}. The SI units of area are m^2. So The SI units of pressure are therefore kg m s$^{-2} \div$ m^2 giving kg m^{-1} s^{-2}.

   (c) momentum = mass × velocity, so SI units are kg m s^{-1}.

2. (a) kJ: the k is kilo = 1000
   ms: the m = 1/1000th

   (b) kJ ms^{-1} is energy per second, ie power.

   (c) Power = energy ÷ time. Any energy equation can be used, such as ½mv^2. The SI Base units are kg (m s^{-1})^2/s = kg m^2 s^{-2} × s^{-1} = kg m^2 s^{-3}.

3. Rearrange to give $h = E\lambda/c$ = joules × metres ÷ velocity = kg m^2 s^{-2} m ÷ m s^{-1}. So the SI base units are kg m^2 s^{-1}.

4. (a) 0.231 m (b) 25000 ÷ 3600 = 6.9 m s^{-1} (c) 3500 kJ

Exercise 2 – *Unit 1.2*

1. (a) Displacement and Velocity are vectors.

   (b) Vector quantities have a magnitude, a unit and a direction. Scalar quantities have only a magnitude and a unit.

2. (a) Horizontal component = 12 cos 35° = 9.8 N

3.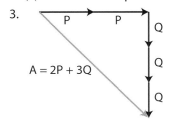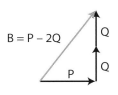

4. (a) 210 sin 55° = 172 N

   (b) Horizontal components are equal in magnitude, but opposite in direction.

   (c) 128 cos 20° = 120 N and 210 cos 55° = 120 N

5. Vertical components: 25 cos 30° = 21.7; 35 cos 60° = 17.5 N
   Vertical resultant of these components = 39.2 N
   Horizontal components: 25 cos 60° (to the right) = 12.5 N; 35 cos 30° (to the left) = 30.3 N. Horizontal resultant of these components = 17.8 N (to the left).
   Resultant = 43 N and the angle = 24° (to 2 sf).

Exercise 3 – *Unit 1.3*

1. Let the distance from A to the cable C be d and **take moments about end A**.
   Sum of Anti–clockwise moments about A = Sum of Clockwise moments about A
   (Note that the weight of S$_1$ does not have a moment about the point A)
   430 × d + 430 × 22 = 120 × 16 + 70 × 24 + 600 × 12
   430 d + 9460 = 1920 + 1680 + 7200
   430d = 1340
   d = 3.1 m

2. The weight of the wheel W, acts vertically downward from the centre of gravity, X. The wheel is in contact with the kerb at E, so there is a normal reaction force R at this point. The wheel is just on the point of moving over the kerb, so it is in equilibrium. This means that the three forces acting on the wheel, the weight, the pulling force of 240 N and the normal force from E must act through the same point. This point is the axle of the wheel, X. Taking moments about the point E means that we ignore the reaction force R as it does not have a moment about this point.

   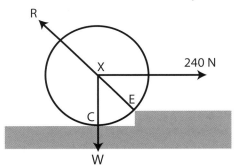

   The perpendicular distance from E to the 240 N force is 0.3 m. The perpendicular distance from E to the line of action of the weight W is 0.4 m (by Pythagoras).
   Taking moments about E: 240 × 0.3 = W × 0.4
   giving W = 180 N

3. (a) The weight of the **diver** and the perpendicular distance from diver to pivot.

   (b) (i) Maximum moment = Weight of diver × maximum distance from diver to pivot.
      Maximum distance to pivot = (4.88 – 1.60) + 0.28 = 3.56 m
      Maximum moment = Fd = 65 × 9.81 × 3.56 = 2270 N m

      (ii) Moment = Weight of diver × distance from diver to pivot
      2270 = 75 × 9.81 × d = 735.75 d
      d = 2270 ÷ 735.75 = 3.09 m
      Distance from diver to central position = 4.88 – 1.60 = 3.28 m
      Pivot must move 3.28 – 3.09 = 0.19 m to the right

4. (a) Take moments about left support:
$30 \times 1 = 18 \times 0.5 + F_y \times 1.5$ giving $F_y = 14$ N
Take moments about right support:
$30 \times 0.5 + 18 \times 2.0 = F_x \times 1.5$ giving $F_x = 34$ N

(b) Take moments about left support, and let weight at left end be F and set $F_y = 0$:
$F \times 0.5 = 30 \times 1$, giving F = 60 N , so weight must be increased by (60 – 18) = 42 N

5. (a) At the point of tipping clockwise moments about support A equal anticlockwise moments about the same point. At the point of tipping the beam is no longer in contact with support B so the reaction force at that point is zero.
CM = ACM: $124 \times 70 = W \times 20$, giving W = 434 N

(b) In vertical equilibrium the upward force = downward forces. Downward force = weight of the beam + gymnast weight = 124 + 434.
So upward force = 558 N.

6. (a) 1. "when an object moves" should read "when an object is in (rotational) equilibrium".

2. It must also be stated that the moments must be taken about the same point.

(b) Take moments about the elbow. CM = ACM
$F_B \times 0.068 = (25 \times 0.19) + (1.5 \times 9.81 \times 0.37)$
$F_B = 150$ N

(c) In equilibrium upward forces = downward forces
$F_B = F_E + 25 + 1.5 \times 9.81$
Re-arranging gives $F_E = F_B - 25 - 1.5 \times 9.81$

(d) $F_E = 150 - 25 - 14.7 = 110.3$ N

7. (a) The required moment is the same with both shaft lengths. $F_{longest} \times 54 = F_{shortest} \times 32$
$F_{longest} \div F_{shortest} = 32 \div 54 = 0.59$
This is a 41% reduction in the required force.

(b) The perpendicular distance between the pivot and the force is = 0.32 cos 34° = 0.27 m
Force = $62 \times 9.81 = 608$ N
Moment = $F \times d = 608 \times 0.27 = 160$ N m

Exercise 4 – *Unit 1.4*

1. (a) 270 kmh⁻¹ = 270 000 metres in 3 600 seconds
= 270 000 m ÷ 3 600 s = 75 m s⁻¹

(b) $a = (v - u) \div t = (75 - 0) / 28 = 2.68$ m s⁻²
$S = \frac{1}{2}(u + v)t = \frac{1}{2}(0 + 75) \times 28 = 1050$ m

2. In last 60.0 metres, constant speed = distance / time
= 60.0 / 4.62 = 12.99 m s⁻¹
In first 40 metres, $S = \frac{1}{2}(u + v)t$
$40 = \frac{1}{2}(0 + 12.99)t$
$t = 80 \div 12.99 = 6.16$ s

Total time = 6.16 + 4.62 = 10.8 s (to 3 significant figures)

3. From t = 0 to t = t_1: non-uniform, decreasing acceleration from zero velocity to a uniform velocity.
From t = t_1: to t = t_2: uniform velocity.
From t = t_2: to t = t_3: uniform deceleration to zero velocity.

4.
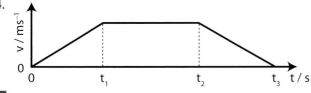

5. (a) speed = 0, acceleration = –9.81 m s⁻²
(b) $v^2 = u^2 + 2as$. So $0^2 = 39.24^2 + 2 \times (-9.81) \times s$ which gives s = 78.48 m
(c) v = u + at. so, 0 = 39.24 + (–9.81)t which gives t = 4 s

6. (a) $v = u + at = 0 + (-9.81) \times 5 = -49.1$ m s⁻¹ where the minus sign shows direction
(b) average speed = $\frac{1}{2}(u + v) = 24.6$ m s⁻¹
(c) s = average speed x time = $24.6 \times 5 = 123$ m

7. (a) $v^2 = u^2 + 2as$. so $0^2 = 12^2 + 2 \times (-9.81) \times s$ which gives s = 7.34 m and hence a total height of $30 + 7.34 \approx 37.3$ m
(b) v = u + at so, 0 = 12 –9.81 × t giving t = 1.22 s
(c) $s = ut + \frac{1}{2}at^2$ so, $-37.3 = 0 + \frac{1}{2} \times (-9.81) \times t^2$ which gives t = 2.76 s
(d) v = u + at = 0 + (–9.81) × 2.76 = 27.1 m s⁻¹

8. (a) velocity = +4.00 m s⁻¹, acceleration = –9.81 m s⁻²
(b) $v^2 = u^2 + 2as$. so $0^2 = 4^2 + 2 \times (-9.81) \times s$
so, s = 0.8 m and hence total height = 22 + 0.8 = 22.8 m
(c) $v^2 = u^2 + 2as = 0^2 + 2 \times (-9.81) \times 22.8 = 447.3$ giving v = –21.2 m s⁻¹
(d) Time to reach max height: t = 4 ÷ 9.81 = 0.41 s
Time to fall from max height to ground = 21.2 ÷ 9.81 = 2.16 s
So total time = 0.41 + 2.16 = 2.57 s

9. (a) Component of g down the slope, ie g sin 30° = 9.81 × 0.5 = 4.91 ms⁻²
(b) Using $v^2 = u^2 - 2as$:
0 = 2.52 – 2 × 4.91 × s, giving s = 0.64 m
(c)
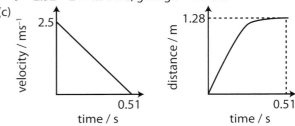

10. (a) Use the equation $s = ut + \frac{1}{2}gt^2$ with u = 0. Re-arrange to $g = 2s \div t^2$. Calculate two values of g and take the average: (2 × 0.4) ÷ 0.29² = 9.5 and (2 × 0.6) ÷ 0.36² = 9.3. Average = 9.4 m s⁻².

(b) A metre rule to measure distance s and electronic timer or light gate linked to a computer to measure the time.

(c) The distance was not measured from base of the sphere. The distance was not read correctly, leading to a parallax error.

11. (a)(i) (200 × 1000) ÷ 3600 = 55.6 m s⁻¹.
(ii) Velocity-time graph shown below
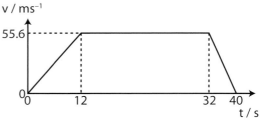

(b) Distance = area between the graph and the time axis.
= ½(55.6 × 12) + 55.6 × 20 + ½(55.6 × 8) = 1670 m

12. (a) $s = \frac{1}{2}(u + v) \times t = \frac{1}{2}(72.0 + 8.5) \times 12 = 483$ m

(b) $v^2 = u^2 + 2as$: $80^2 = 0 + 2 \times 0.96 \times s$ giving s = 3333 m
This is the required distance, hence the runway is short by 3333 − 2780 = 553 m.

Exercise 5 – *Unit 1.5*

1. (a) Horizontal velocity component
= 25 ÷ 1.4 = 18 m s⁻¹
Initial vertical velocity component v = u + at so,
0 = u − 9.81 × 1.4, giving u = 14 m s⁻¹
(b)(i) By Pythagoras, $v = \sqrt{(17.9^2 + 13.7^2)} = 23$ m s⁻¹
(ii) θ = tan⁻¹(13.7÷17.9) = 37°
(iii) s = ½ (u+v)t = ½ (13.7+0) × 1.4 = 9.6 m

2. (a) Initial vertical velocity = 9.3 sin 45 = 6.6 m s⁻¹
(b) For the vertical motion, the displacement s is −0.6 m, so, using $v^2 = u^2 + 2as = 6.58^2 + 2(−9.81)(−0.6) = 55.07$
$v = \sqrt{55.07} = 7.4$ m s⁻¹
(c) v = u + at, so −7.4 = 6.58 − 9.81t
giving t = (6.58 + 7.4) ÷ 9.81 = 1.4 s

3. (a) The horizontal component of the initial velocity
= u cos 40°
The vertical component of the initial velocity = u sin 40°
So horizontal time T_H = 36 ÷ u cos 40°
The time to reach maximum height = u sin 40° ÷ g
Remember at maximum height the vertical velocity is momentarily zero (use v = u − gt). It takes the same time to descend from the maximum height.
So vertical time T_V = 2u sin 40° ÷ g
(b) T_H = T_V so: 36 ÷ u cos 40° = 2u sin 4° ÷ g
Re-arranging gives: u² = (36 × 9.81) ÷ (2 × 0.643 × 0.766)
u = 18.9 m s⁻¹

4. (a) u sin 30° = 30 m s⁻¹ (from the graph) so the initial velocity u = 60 m s⁻¹.
(b) Time to reach max height t = u sin 30° ÷ g
= 30 ÷ 9.81 = 3.06 s
Max height = ½(30 + 0) × 3.06 = 46 m
(c) Range = horizontal velocity × 2 × time to reach max height
= 60 cos 30° × 2 × 3.06 = 320 m

5. (a) At max height vertical velocity = 0 using v = u − gt
0 = 13.5 sin 40° − 9.81t, giving t = 0.89 s
(b) Height above launch = ½ (13.5 sin 40° + 0) × 0.89
= 3.86 m. So height above ground = 3.86 + 2.0 = 5.9 m.
(c) Total time in the air = time to reach max height + time to fall vertically from rest from the max height.
For the time to fall vertically from rest from the max height use s = ut + ½ gt²:
5.84 = 0 + 1.2 9.81 × t² giving t = 1.09 s
Total time = 0.89 + 1.09 = 1.98 s
Horizontal distance = horizontal velocity × total time in the air = 13.5 cos 40° × 1.98 = 21 m

Exercise 6 – *Unit 1.6*

1. a = F÷m = −1200 ÷ 1800 = −0.667 m s⁻¹
v = u + at, so 0 = 16.7 − 0.667t
so: t = 16.7 ÷ 0.667 = 25 s

2. (a) Opposing forces = 2.3×10⁶ × 0.6 = 1.38×10⁶ N
Resultant force = ma = 2.3×10⁶ × 0.2 = 4.6×10⁵ N
Driving force = Resultant force + Opposing forces
= 1.38×10⁶ + 4.6×10⁵ = 1.8×10⁶ N

(b) Friction is reduced between wheel and track so resultant forward force decreases. So the acceleration decreases.

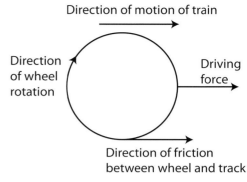

3. (a) See text.
(b) (i) The brick is stationary, so by Newton's First Law the normal contact force exerted by the ground on the brick equals the weight of the brick, ie A = B.
(ii) The downwards force exerted by the brick on the ground is equal to the normal contact force exerted by the ground on the brick, ie C = A.
The gravitational attraction of the brick on the Earth equals the weight of the brick, ie D = B.

4. (a) The rate of change of momentum (or the acceleration of a constant mass) is proportional to the resultant force and takes place in the same direction as the resultant force.
(b) (i) The reaction force when the lift is stationary is equal to her weight. mg = 579 N
(ii) The reaction force is a maximum when the acceleration of the lift and the weight of the passenger are in opposite directions. In this case it happens when the lift decelerates as it approaches the ground floor. Reaction = mg + ma
= 579 + (59 × 2.2) = 709 N
(c) If lift falls freely or accelerates downwards at 9.81 m s⁻² then the reaction from the floor will be zero.

5. F_resultant = ma
F_resultant = driving force − friction − component of the car's weight acting down the slope
= 8000 − 200 − (1480 × 9.81 × sin 12°) = 1480a
a = 3.2 m s⁻²

Exercise 7 – *Unit 1.7*

1. (a) p = mv = 0.012 × 300 = 3.6 Ns
(b) Momentum before collision = momentum after collision, so: 3.60 = 3.012v giving v = 1.2 m s⁻¹.
(c) KE = ½mv² = ½ × 3.012 × 1.195² = 2.2 J
(d) mgh = KE, so 3.012 × 9.81 × h = 2.15, giving h = 7.3 cm

2. (a) P = mv = 1200 × 6.0 = 7200 kg m s⁻¹
(b) Momentum before collision = momentum after collision, so: 7200 = (1200 + M) × 2.0
giving M = 2400 kg
(c) Inelastic because kinetic energy is not conserved.
Initial KE = ½ × 1200 × 6² = 21600 J
Final KE = ½ × (1200 + 1800) × 2² = 6000 J

3. (a) The initial momentum is zero and so in order to conserve momentum the pieces move off in opposite

directions. Momentum before explosion = Momentum after explosion, so: $0 = MV - mv$. Since M is greater than m, then v must be greater than V.

(b) Using the principle of conservation of momentum:
$7.26 \times 8.15 = 1.47 \times 13.32 + 7.26 \times V$
giving $V = 5.45$ m s^{-1}.

(c) In an inelastic collision kinetic energy is not conserved.

4. (a) $P = mv = 6.64 \times 10^{-27} \times 1.6 \times 10^4 = 1.06 \times 10^{-22}$ kg m s^{-1}

(b) Initial momentum = 0. The lead nucleus and the alpha particle move way in opposite directions with equal amounts of momentum, $p_{lead} = p_{alpha}$. Therefore:
$206 \times V_{lead} = 4 \times V_{alpha}$ so, $206 \times V_{lead} = 4 \times 1.6 \times 10^4$
giving $V_{lead} = 311$ m s^{-1}.

(c) This decay is inelastic since kinetic energy is not conserved. The polonium nucleus is stationary, ie it has zero kinetic energy, before the decay. The resulting lead nucleus and alpha particle both have kinetic energy.

5. (a) Vertically downwards. The ejection seat moves upwards and to conserve momentum the aircraft moves in the opposite direction.

(b) Momentum of the ejection seat = momentum of aircraft, so: $200 \times 180 = 8920 \times V$, giving $V = 4.0$ m s^{-1}.

(c) Kinetic energy is not conserved. The aircraft containing the ejection seat initially has zero kinetic energy but after ejection both have kinetic energy.

6. Applying the Principle of Conservation of momentum we have: $110 \times 2 - 440 = -220$ (the minus being because the players are moving in opposite directions) giving:
$(110 + M) \times V = 220$ (Equation 1)
$KE = \frac{1}{2}mv^2$, giving: $\frac{1}{2}(110 + M) V^2 = 115$ (Equation 2)
Substituting Equation 1 into Equation 2 gives
$\frac{1}{2} \times 220 \times V = 115$, so $V = 1.05$ m s^{-1}.
After the collision the two players move in the opposite direction to the player of mass 110 kg.

Exercise 8 – *Unit 1.8*

1. Additional force to be overcome due to hill
$= mg \sin \theta = 1200 \times g \times \sin 5.7° = 1169.19$ N
Additional power required
$= Fv = 1169.19 \times 30 = 35\ 076$ W ≈ 35 kW
Total power required $= 60 + 35 = 95$ kW (to 2 sf)

2. $P = Fv$, so $2400 = 8000 \times v$
$v = 0.3$ m s^{-1}

3. (a) Kinetic energy: energy possessed by body because of its motion. Potential energy: energy possessed by body because of its height above the ground.

(b) Straight line graph through (0,3) and (3,0). Energy conservation applied because the total energy at all points on the graph $(E_k + E_p) = 3$ J.

(c) (i) Total energy $= mg\Delta h + \frac{1}{2}mv^2$
$= (75 \times 9.81 \times 1.60) + (\frac{1}{2} \times 75 \times 0.80^2)$
$= 1200$ J (to 2 sf)

(ii) $E_{total} = \frac{1}{2}mv^2 = \frac{1}{2} \times 75 \times v^2 = 1200$

$v = \sqrt{1200 \div 37.5} = 5.7$ ms^{-1} (to 2 sf)

4. (a) useful output power = efficiency $\times$ total input power
$= 0.04 \times 60 = 2.4$ W

(b) efficiency $= \dfrac{\text{useful power output}}{\text{total power input}} = \dfrac{2.4}{12} = 0.2 = 20\%$

5. (a) Work done per second = Power = Fv
$= 50\ 000 \times 30 = 1\ 500\ 000$ W = 1.5 MW
So train does 1.5 MJ every second

(b) $P = kv^3$, so $1.5 \times 10^6 = k \times 30^3$ giving $k = 55.556$
At 40 m s^{-1}, $P = 56.5556 \times 40^3 = 3.56$ MW ≈ 3.6 MW (to 2 sf)

6. (a) KE at bottom $= 0.85 \times$ GPE at top
$= 0.85 \times (90 \times 9.81 \times 8) = 6004$ J
$KE = \frac{1}{2}mv^2 = 6004 = \frac{1}{2} \times 90 \times v^2$ so
$v = \sqrt{6004 \div 45} = 11.55 \approx 12$ ms^{-1} (to 2 sf)

(b) KE at top of second hill $= \frac{1}{2}mv^2 = \frac{1}{2} \times 90 \times 8.9^2 = 3564.5$ J
Change in PE going up second hill $= mg\Delta h$
$= 90 \times 9.81 \times 6 = 5297$ J
Work done $= (PE + KE)_{at\ top} - KE_{at\ bottom}$
$= (5297 + 3564.5) - 6004 = 2858$ J
Work done ≈ 2900 J (to 2 sf)

7. 30 MW $= 3.0 \times 10^6$ Js$^{-1} = m \times 9.81 \times 6$ giving $m = 509\ 684$ kgs^{-1}
The answer to 2 significant figures is 510 000 kgs^{-1}

8. (a) Work done = force $\times$ distance moved:
$19.2 = F \times 6.59$, giving $F = 2.91$ N

(b) Opposing force on swept area is 88% that of unswept area. Total KE = KE on unswept area + KE on swept area. Using work = force $\times$ distance:
$(2.91 \times 3.59) + (0.88 \times 2.91 \times 3.0)$
giving KE $= 18.1$ J

9. (a) The product of a force and distance moved in the direction of the force.

(b) Steady speed means the horizontal component of the pulling force = frictional force.
Friction $= 240 \sin 28°$ or $240 \cos 62° = 110$ N (to 2 sf)

(c) Work = force $\times$ distance $= 110 \times 36 = 3960$ J

10. GPE $= mgh = 78 \times 9.81 \times 19.5 = 14\ 921$ J
KE $= 0.9$ GPE $= 13\ 429$ J
KE $= \frac{1}{2}mv^2$, so: $13\ 429 = \frac{1}{2} \times 78 \times v^2$ giving $v = 18.6$ m s^{-1}
Giving the answer to 2 sf we have 19 m s^{-1}.

11. Work done by the resistive force = initial kinetic energy of the javelin. $F_{resistive} \times 0.065 = 245$, giving $F_{resistive} = 3769$ N.

Exercise 9 – *Unit 1.8*

1. E.m.f. is the potential difference across the output terminals of the source when no current is being drawn from it. Terminal potential difference is the voltage across the terminals when a current is being drawn from the source.

2. Charge $Q = It = 1 \times 20 \times 3600 = 72\ 000$ C = 72 kC
Energy $= QV = 72\ 000 \times 12 = 864\ 000$ J = 864 kJ

3. (a) (i) $\Delta Q = I \times \Delta t = 0.4 \times (3 \times 60 \times 60) = 4320$ C
(ii) Number of electrons = total charge $\div$ charge on 1 electron $= 4320 \div 1.6 \times 10^{-19} = 2.7 \times 10^{22}$

(b) At the instant the potential difference is applied, all the free electrons acquire a drift speed at all points in the conducting cable.

4. (a) An electric current is the electric charge passing a fixed point in one second.

(b) $I = \Delta Q \div \Delta t = (5 \times 10^{20} \times 1.6 \times 10^{-19}) \div 25 = 3.2$ A

(c) (i) $v = s \div t$, so $t = s \div v = 0.45 \div 8 \times 10^6 = 5.6 \times 10^{-8}$ s

(ii) $Q = It = 1.85 \times 10^{-3} \times 5.625 \times 10^8 = 1.04 \times 10^{-10}$ C
Number of electrons = total charge $\div$ charge on 1 electron $= 1.04 \times 10^{-10} \div 1.6 \times 10^{-19} = 6.5 \times 10^8$

5. (a) (i) Electric current is rate of flow of charge.

    (ii) Potential difference is the energy transferred from electrical (to other forms) per coulomb.

   (b) $Q = It = 0.025 \times 120 = 3$ C

   (c) Energy $= QV = 3 \times 6 = 18$ J

6. (a) (i) 12 mA = 12 mC per second i.e. 12×10^{-3} Cs^{-1}

    (ii) Energy $= QV = 12 \times 10^{-3} \times 6.3 = 76 \times 10^{-3}$ J

   (b) $Q = It = 12 \times 10^{-3} \times 90 = 1.08$ C.
   Each electron has a charge of 1.6×10^{-19} C, so number of electrons $= 1.08 \div 1.6 \times 10^{-19} = 6.8 \times 10^{18}$.

7. (a) $I = Q \div t = (4.11 \times 10^{21} \times 1.6 \times 10^{-19}) \div 126 = 5.22$ A

   (b) $P = IV = 5.22 \times 230 = 1.2$ kW

   (c) Energy $=$ power $\times$ time $= 1200 \times 126 = 1.51 \times 10^5$ J

Exercise 10 – *Unit 1.10*

1. (a) (i) 18 Ω are in parallel with 6 Ω to give a total resistance of $(18 \times 6) \div (18 + 6) = 4.5$ Ω

    (ii) $I_1 = I_2 + I_3$

    (iii) The p.d. across the network $= IR = 6 \times 4.5 = 27$ V $=$ p.d. between X and Y. So $I_3 = V \div R = 27 \div 6 = 4.5$ A

   (b) Since the pd across the network is 10 V and all resistors are identical, opposite ends of the meter are at the same potential of 5 V. So there is no potential difference across the ammeter, so no current flows in the ammeter.

2. (a) First parallel network has resistance $(20 \times 60) \div (20 + 60) = 15$ Ω
   Second parallel network has resistance $(24 \times 48) \div (24 + 48) = 16$ Ω
   Total resistance between A and B $= 15 + 16 = 31$ Ω

   (b) $R_{circuit} = V_{battery} \div I_{battery} = 12 \div 0.3 = 40$ Ω

   (c) So resistor R has resistance $40 - 31 = 9$ Ω

   (d) Voltage across second parallel combination $= IR = 0.3 \times 16 = 4.8$ V.
   Current in 48 Ω resistor $= V \div R = 4.8 \div 48 = 0.1$ A = 100 mA

3. (a) The electrical resistivity of a material is defined as numerically equal to the resistance of a sample of the material 1 m long and of cross sectional area 1 m^2.

   (b) (i) $\rho = RA \div L = 9.0 \times \pi \times (1 \times 10^{-4})^2 \div 15 = 1.88 \times 10^{-8}$ Ωm

    (ii) Since the wire in the coils have the same length and cross section area, the difference in their resistance depends only on their resistivity. Hence the resistance of the wire in coil B is 30 times that of the wire in coil A. So the resistance of the wire in coil B is $30 \times 9 = 270$ Ω.

   (c) (i) For the heating element use wire from coil B. To repair the electrical connection use wire from coil A.

    (ii) From Joule's Law: $P = I^2R$
    We generally require a connection not to become hot, so we use a wire of minimum resistance (coil A). But a heater is designed to produce heat and hence it requires a wire of greater resistance (coil B)

4. (a) $R = V \div I = 0.12 \div 3.5 = 0.034$ Ω
   $\rho = RA \div L = 0.034 \times \pi(0.56 \times 10^{-3})^2 \div 2 = 1.67 \times 10^{-8}$ Ωm

   (b) Length and material are unchanged, but the radius is halved. So the area is decreased by a factor of 4

(since $A = \pi r^2$). So resistance increases by a factor of 4 (since $R = \rho L \div A$). Resistivity does not change because resistivity is a property of the material, not the wire.

5. (a) (i) Resistance $= V \div I = 14.2 \div 8.4 = 1.7$ Ω

    (ii) Power $= V \times I = 14.2 \times 8.4 = 119$ W

   (b) The total resistance of N equal resistors R, arranged in parallel, is R ÷ N.
   Resistance of 1 strip $= R \times N = 1.69 \times 6 = 10.1$ Ω

   (c) If one strip open circuited, the whole heating element would fail and supply no heat. A parallel arrangement means there is less current in each strip; a series arrangement would require a higher voltage source to deliver the same current to each strip as the parallel arrangement.

6. (a) Going clockwise from X to Y we have two 6 Ω resistors in series giving 12 Ω. Going anticlockwise from X to Y we also have two 6 Ω resistors in series giving 12 Ω. There are therefore two 12 Ω resistors in parallel, giving a total of $12 \div 2 = 6$ Ω.

   (b) Going clockwise from X to Y we have two 6 Ω resistors in series giving 12 Ω. In parallel with this 12 Ω there is now an additional 6 Ω resistor, giving a combined resistance of $(6 \times 12) \div (6 + 12) = 4$ Ω. This 4 Ω is in series with the 6 Ω connecting Y to Z to give 10 Ω. The 10 Ω is in parallel with the 6 Ω between X and Z, giving a total resistance of $(6 \times 10) \div (6 + 10) = 60 \div 16 = 3.75$ Ω.

7. (a) For an ohmic conductor at constant temperature current and potential difference are proportional.

   (b) (i) Resistance between X and Y $= R = V \div I = 6 \div 0.002 = 3000$ Ω.

    (ii) Two Rs in series equals 2R. This 2R in parallel with 1R gives 2R/3, which is then in series with 1R giving a total of 5R/3 = 3000, giving R = 1800 Ω.

8. (a) Resistance of 6 Ω and 2 Ω in parallel = 1.5 Ω. Resistance of remaining network of 4 resistors $= (12.5 - 6 - 1.5) = 5$ Ω. This is equivalent to two 10 Ω resistors in parallel. The resistance $1 \Omega + 5 \Omega + R = 10$ Ω, so R = 4 Ω.

   (b) Current drawn from battery $= V \div R = 15 \div 12.5 = 1.2$ A. For the resistance of 6 Ω and 2 Ω in parallel the current divides in the ratio 1 to 3. So a current = 0.9 A passes through the 2 Ω resistor. Power $= I^2R = 0.9^2 \times 2 = 1.62$ W.

9. (a) Resistance $R = \rho l \div A$, $\rho = RA \div l$. The area of cross-section $A = \pi d^2 \div 4$. This gives $\rho = R \pi d^2 \div 4l$

   (b) Gradient of graph $I \div R = 0.1$ so $R \div l = 10$. The diameter of the nichrome wire is calculated as 0.36 mm. The diameter of the aluminium wire is calculated to be 0.06 mm. This is too small, so the wire is nichrome.

   (c) Use low current or low voltage so that the dangers of a hot wire are avoided.

10. (a) $R = \rho l \div A$, so $A = 1.45 \times 10^{-6} \times (25.4 \times 10^{-3}) \div 0.19 = 1.94 \times 10^{-7} = \pi d^2 \div 4$ giving d = 0.5 mm.

   (b) Make the wire thinner. A is smaller because the diameter is smaller. If I decreases, R increases.

11. Although increasing the temperature increases the number of collisions between the free moving electrons and the atoms, the temperature increase results in more of the loosely bound electrons escaping from their atoms

and becoming free to move. The second effect more than compensates for the first.

Exercise 11 – *Unit 1.11*

1. (a) The pd across the terminals of the source falls as increasing current is drawn from it, resulting in energy being dissipated as heat in the source of the e.m.f.

   (b) (i) The gradient of the graph is –r, where r is the internal resistance of the cell.

   (ii) Extrapolate the graph to where it cuts the vertical axis. The y-intercept is the e.m.f..

   (c) E = V + Ir and E = 10 volts (terminal p.d. when no current is drawn).
   I = V ÷ R = 9.5 ÷ 2 = 4.75 A
   E = V + Ir, so 10 = 9.5 + 4.75r
   r = 0.5 ÷ 4.75 = 0.1 Ω

2. (a) The emf can be regarded as the reading on a voltmeter connected across the terminals of the battery when no current is drawn (ie open circuit). The terminal potential difference is the reading on the same voltmeter when current is drawn from the battery.

   (b) (i) Internal resistance is the opposition to current flow through the electrical power source due to the resistance of components in source (or resistance of chemicals in battery).

   (ii) See text (and refer also to part (b) of the previous question).

3. (a) Diagram similar to that shown below with ammeter, voltmeter and variable resistor symbols correct, and the components in the correct positions.

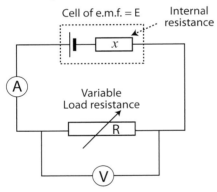

   (b) Quantities are current and terminal voltage.

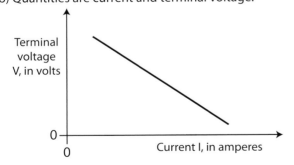

   (c) The magnitude of the gradient of the line gives the internal resistance.

4. (a) Current through resistor = V_2 ÷ R = 5.33 ÷ 16.4 = 0.325 A
   Lost volts = 6.52 – 5.33 = 1.19 V
   Total internal resistance = V ÷ I = 1.19 ÷ 0.325 = 3.66 Ω
   Average internal resistance per cell = 0.92 Ω

   (b) V_1 would be higher since the connecting wires also have a small resistance.

Exercise 12 – *Unit 1.12*

1. (a) V_{out} in bright light = (10 × 15) ÷ (300 + 10) = 0.48 V

   (b) The voltage across the motor must be 6 V, so:
   6 = (15 × R) ÷ (300 + R) (where R is the resistance of the 250 Ω resistor and motor) giving R = 200 Ω. To find R_{motor}: 1/200 = 1/R_{motor} + 1/250 giving R_{motor} = 1000 Ω.

2. (a) Resistance of the voltmeter and R_1 in parallel = 10 Ωk
   V_{out} = (12 × 10) ÷ (20 + 10) = 4 V

   (b) Resistance of the new voltmeter and R_1 in parallel = 19.6 kΩ. This combined resistance is almost equal to the resistance when no voltmeter is connected to the circuit, so the output voltage is the same.

3. (a) R_1 = the resistance of the thermistor. So:
   6 = (14 × R_1) ÷ (R_1 + 3180) giving R_1 = 2385 Ω
   Give the answer as 2400 Ω since 2 sig figs is asked for.

   (b) The combined resistance of the thermistor and heater in parallel is 960 Ω. Let R = the value of the variable resistor required to give an output voltage of 6V:
   V_{out} = 6 = (14 × 960) ÷ (960 + R) giving R = 1280 Ω.

4. (a) (i) In bright light the LDR has a resistance of 500 Ω. So R_1 = 500 Ω.
   $$V_{out} = \frac{R_1}{R_1+R_2} \times V_{IN} = \frac{500}{(500+10000)} \times 12 = 0.6\ V$$

   (ii) In the dark, the LDR has a resistance of 100 000 Ω. So R_1 = 200 000 Ω.
   $$V_{out} = \frac{R_1}{R_1+R_2} \times V_{IN} = \frac{200000}{(200000+10000)} \times 12 = 11.4\ V$$
   Since V_{out} > 10 V, the lamp will light (in the dark).

   (b) If the positions of the LDR and the fixed resistor are swapped, and the output p.d. is now across the fixed resistor, then: In the dark:
   $$V_{OUT} = \frac{R_1}{R_1+R_2} \times V_{IN} = \frac{10000}{(200000+10000)} \times 12 = 0.56\ V$$
   and the lamp will be OFF.
   In bright light:
   $$V_{OUT} = \frac{R_1}{R_1+R_2} \times V_{IN} = \frac{10000}{(500+10000)} \times 12 = 11.4\ V$$
   and the lamp will be ON. So the automatic light will work in reverse, lighting up when bright and going off when dark.

Exercise 13 – *Unit 2.1*

1. (a) Gamma rays, X-rays, Ultra-violet, Visible, Infra-red, Microwaves, Radio

   (b) 400 nm (violet) – 700 nm (red)

   (c) An electric and a magnetic field oscillate at right angles to each other, and both oscillate at right angles to the direction in which the wave is moving. In the following diagram, the dark wave is the electric wave, the lighter wave is the magnetic wave.

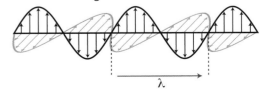

(d) Can travel through a vacuum or have associated oscillating electric and magnetic fields.

2. (a) $T = \dfrac{1}{f} = \dfrac{1}{5} = 0.2$ s

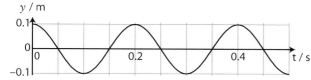

(b) $\lambda = \dfrac{v}{f} = \dfrac{20}{5} = 4$ m

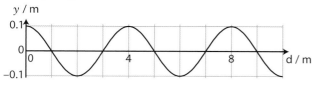

(c) $\phi = \dfrac{0.8}{4} \times 360° = 72°$

3. Period $T = 0.4$ s. Frequency $= \dfrac{1}{T} = \dfrac{1}{0.4} = 2.5$ Hz

4. (a) (i)

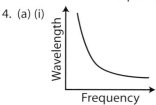

(ii) A graph showing a horizontal line parallel to x-axis at 3×10^8 m s^{-1}.

(b) Since $f = c\lambda^{-1}$, the graph is a straight line through (0,0) with a gradient equal to the speed of light, c. So the gradient is 3×10^8 m s^{-1}.

Exercise 14 – Unit 2.2

1. Tabulate corresponding values of sin i and sin r.
i = angle of incidence and r = angle of refraction.
Plot a graph of sin i (y-axis) against sin r (x-axis).
Determine the gradient of the best fit straight line.
The gradient = sin i ÷ sin r = refractive index.

2. (a) Meeting the glass-air boundary at the critical means that the ray of light travelling from the glass into the air will be refracted into the air at an angle of 90° to the normal. This means it will travel along the glass-air boundary.

(b) If the angle was greater than the critical angle the ray of light would undergo total internal reflection. It will be reflected back into glass with the angle of reflection being the same as the angle of incidence.

(c) sin C = 1/refractive index = 1÷1.39 = 0.7194. C = 46°.

(d) Look at the angles on the diagram:

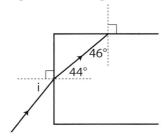

$n = \sin i \div \sin r$
$1.39 = \sin i \div \sin 44$
$\sin i = 1.39 \times 0.6947 = 0.9656$
$i = 74.9°$

3. The diagram shows the ray of light.

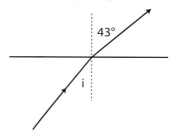

Use Snell's law to calculate i
$\sin i \div \sin 43 = 1 \div 1.38 = 0.725$.
Remember the light is moving from glass into air so $_{glass}n_{air}$
$= 1 \div {}_{air}n_{glass}$
$\sin i = 0.4944$ giving $i = 29.6°$.
To undergo total internal the angle i must just exceed the critical angle C.

To calculate C we use $\sin C = 1 \div 1.38 = 0.725$ giving a value for $C = 46.5°$.
The required increase in the value of the angle i is $46.5° - 29.6° = 16.9°$.

4. (a) Speed = distance÷time = 1200 m ÷ 5880×10^{-9}
$= 2.04 \times 10^8$ m s^{-1}.

(b) Refractive index of a material = velocity of light in vacuum ÷ velocity of light in the material.
The velocity of light in a vacuum (data sheet) $= 3.0 \times 10^8$ m s^{-1}.
Refractive index $= 3 \times 10^8 \div 2.04 \times 10^8 = 1.47$

(c) $\sin C = 1 \div 1.47 = 0.6803$ giving $C = 42.9°$.

5. (a) The refractive index does not change gradually at the core-cladding boundary, rather it changes suddenly like a step. Moreover, the refractive index inside the cote is constant and greater than that of the cladding.

(b) $_{cladding}n_{core} = 1 \div \sin C = 1 \div \sin(60) = 1.15$

(c) Speed of light in core = $3 \times 10^8 \div 1.6 = 1.88 \times 10^8$ m s^{-1}
Time to travel 500 m along axial path
$= 500 \div 1.88 \times 10^8 = 2.66$ μs
Suppose axial mode distance corresponding to a single TIR is x, then the corresponding distance at the critical angle is $x \div \sin C = x \div \sin 60° = 1.15x$
Distance travelled in this mode = $500 \times 1.15 = 575$ m
Time to travel this distance = $575 \div 1.88 \times 10^8$
$= 3.06$ μs. Time difference = $3.06 = 2.66 = 0.40$ μs.

(d)(i) The engineer would want a cladding with a larger refractive index so that C is larger and dispersion is reduced.

(ii) Increase the critical angle

(iii) Reduces dispersion by reducing higher modes of propagation.

6. (a) $\sin 45° \div \sin r = 1.52$ giving $r = 27.7°$.

(b) Use the hint given $r = 60° - 27.7° = 32.3°$.

(c) The critical angle for the material of the prism is found by: $\sin C = 1 \div 1.52 = 0.6579$, $C = 41.1°$

The angle of incidence in the glass, namely the angle r,

is only 32.3°. This is less than the critical angle so total internal reflection does not happen.

(d) sin 32.3° ÷ sin θ = 1 ÷ 1.52
(the light is travelling from glass to air - see solution to question 3) sin θ = 0.8015, give a value for θ = 53.3°.

Exercise 15 – Unit 2.3

1. The lens shown is a diverging or concave lens.

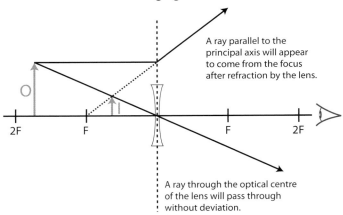

A ray parallel to the principal axis will appear to come from the focus after refraction by the lens.

A ray through the optical centre of the lens will pass through without deviation.

2. (a) The diagram of the apparatus.

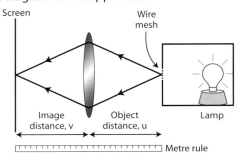

(b) Place the lens between the lamp house and the screen. Adjust the position of the screen until a sharp image is seen on the screen.

(c) Measure the distance from the lens to the wire mesh (object distance u). Measure the distance from the lens to the screen (image distance v). Repeat this process for five or six different object distances.

(d) Plot a graph of 1/u against 1/v. Draw the best fit line through the points. Extend the line until it cuts both axes. The intercepts on each axes is equal to 1/f, f is the focal length. Take an average value of 1/f and calculate f. (see the graph under 'Measuring Object and Image Distances' in text).

3. (a) The magnification, m = v÷u = 6. Thus v = 6u.

Use the lens formula $\frac{1}{u}+\frac{1}{v}=\frac{1}{f}$

Since the image is virtual, image distance is negative:

$\frac{1}{u}-\frac{1}{6u}=\frac{1}{10}$ giving $\frac{(6-1)}{6u}=\frac{1}{10}$, so $\frac{6u}{5}=10$

Thus u = 8.33 cm

(b) The image distance v = 6u = 6 × 8.33 = 50 cm.

4. (a)

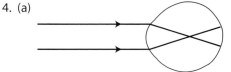

A person who suffers from myopia (short sight) is unable to

see distant objects sharply. They cannot make the lens thin enough to view distant objects. This causes the light from distant objects to converge towards a point in front of the retina. The image seen by the person is blurred.

(b)

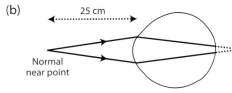

25 cm

Normal near point

A person who suffers from hyper-metropia (long sight) sees distant objects clearly but does not see near objects clearly. An object held at the normal near point distance of 25 cm will not be seen clearly. The rays of light from the object are not bent sufficiently to form an image on the retina. The rays converge behind the retina.

5. (a) Long sight (hypermetropia).

(b) The eyeball is too short – cornea is not curved enough.

(c) The student's near point is 80 cm. We must find the power of the converging lens that will cause an object at the normal near point (25 cm) to give a virtual image at 80 cm.

$\frac{1}{u}+\frac{1}{v}=\frac{1}{f}$, so: $\frac{1}{f}=\frac{1}{25}-\frac{1}{80}=\frac{11}{400}=36.36$ cm = 0.36 m

(d) (i) We need to locate the position of the real object which will give an image at an infinite distance from the lens. But this is the focal length of the lens. So the student's far point is 36.4 cm from his eye.

(ii) The student's range of vision with this lens is 25 cm to 36.4 cm, so an object at 5 m will appear blurred.

6. (a) When unaccommodated the object distance is infinite, so $\frac{1}{u}=0$. So the image distance is the focal length of the eye lens: f = $\frac{1}{P}=\frac{1}{50}=0.02$ m = 2 cm

(b) $\frac{1}{u}+\frac{1}{v}=\frac{1}{f}$, so: $\frac{1}{25}+\frac{1}{2}=\frac{1}{f}$, giving $\frac{1}{f}=\frac{27}{50}$ cm^{-1} = 54 m^{-1}

P = $\frac{1}{f}$ = 54 D. So the increase in power = 54 – 50 = +4 D.

Exercise 16 – Unit 2.4

1. (a) (i) and (ii)

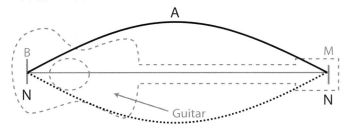

(b) (i) For the first mode of vibration the length of string is ½ λ. The wavelength λ is therefore 1.68 m.

(ii) A frequency of 328 Hz is 4 times the lowest frequency of 82 Hz. The new wavelength will therefore be ¼ of 1.68 m = 0.42 m. The length of string BX will be ½ of this and equal to 0.21 m.

(c) The point F is ⅔ of the way along the string: there will be a node at this point as well as nodes at B and M.

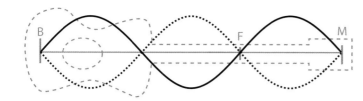

2. (a) The size of the gap is nearly the same as the wavelength of waves so the amount of diffraction will be high.

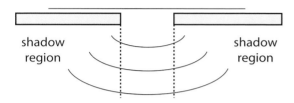

shadow region   shadow region

(b) As the wavelength of the sound increases the amount of diffraction will increase and the sound will spread more into the shadow zone. This means the shadow zone will get smaller.

3. (a) In general, the phase difference for coherent light is constant, but not necessarily zero. However for laser light, the phase difference between S and T is zero.

(b) Constructive interference arises when the path difference between the interfering beams is $n\lambda$ where n is an integer or zero. Destructive interference arises when the path difference between the interfering beams is $(n + \frac{1}{2})\lambda$ where n is an integer or zero. As we move away from the axis of symmetry, n increases, so we get a series of bright and dark bands.

(c)(i) $y = 24.6 \div 6 = 4.1$ mm (notice that there are 6 fringe widths, not 7)

(ii) $\lambda = ay \div d = (0.6 \times 10^{-3} \times 4.1 \times 10^{-3}) \div 3.9$
$= 6.31 \times 10^{-7}$ m $= 631$ nm

(iii) Red

4. The fringe width is directly proportional to the wavelength. So the new fringe width $= (600 \times 500) \div 300$ μm $= 750$ μm. So the increase in the separation of the fringes is $750 - 500 = 250$ μm.

5. With the sodium lamp the fringes are often quite faint. Reducing the background light enhances their visibility by increasing contrast. With a laser an important safety precaution is keep the pupil of the eye as small as possible to minimise possible damage caused be reflected laser light. Reducing the ambient light causes the size of the pupil to increase and this would increase the harm caused to the retina if laser light entered the eye.

6. (a) $d = 1 \div 250 = 4 \times 10^{-3}$ mm $= 4 \times 10^{-6}$ m

(b) θ = angle between diffracted ray and line of symmetry $= 36.8° \div 2 = 18.4°$

(c) $d \sin\theta = n\lambda$, so $\lambda = d \sin\theta \div n = (4 \times 10^{-6} \times \sin 18.4°) \div 2$
$= 6.31 \times 10^{-7}$ m $= 631$ nm

(d) The highest possible angle of diffraction is 90°, for which $\sin\theta = 1$. For the highest order, $n = d \sin\theta \div n$

$= (4 \times 10^{-6} \times 1) \div 6.31 \times 10^{-7} = 6.34$. But since the order is an integer, highest order observed must be 6.

Exercise 17 – Unit 2.5

1. (a) $KE_{max} = \Phi - (hc \div \lambda)$
$= 3.7 \times 10^{-19} - (6.63 \times 10^{-34} \times 3 \times 10^8 \div 4.76 \times 10^{-7})$
$= 4.79 \times 10^{-20}$ J. Question asked for value in eV, so:
$KE_{max} = 4.79 \times 10^{-20} \div 1.6 \times 10^{-19}$ eV $= 0.3$ eV

(b) 0.3 V

2. (a) Frequency of incident light $= 3 \times 10^8 \div 4.50 \times 10^{-7}$
$= 6.67 \times 10^{14}$ Hz
Threshold frequency $= \Phi \div h = 4.3 \times 10^{-19} \div 6.63 \times 10^{-34}$
$= 6.49 \times 10^{14}$ Hz. Since frequency of incident light > threshold frequency, photoelectric emission occurs.

(b) Photoelectric emission still occurs because incident frequency has increased. But, number of photoelectrons emitted per second is independent of frequency. So there is no change to the number of electrons emitted per second from the surface.

3. (a) $E = hc \div \lambda = (6.63 \times 10^{-34} \times 3 \times 10^8) \div 658 \times 10^{-9}$
$= 3.023 \times 10^{-19}$ J. Convert to eV, so:
$E = 3.023 \times 10^{-19} \div 1.6 \times 10^{-19} = 1.89$ eV
The electron therefore moves from the −1.51 eV energy level to the −3.40 eV energy level.

(b) Most energetic transition has energy 1.51 eV
$= 1.51 \times 1.6 \times 10^{-19}$ J $= 2.42 \times 10^{-19}$ J
$\lambda = hc \div E = (6.63 \times 10^{-34} \times 3 \times 10^8) \div 2.42 \times 10^{-19}$
$= 822$ nm, which is greater than 700 nm and in the infrared part of the electromagnetic spectrum.

4. (a) Electrons in atoms orbit nuclei in shells. If a vacancy arises in a shell of low energy, an electron in a higher energy level may spontaneously and randomly move into the vacancy and in so doing cause the emission of a quantum of light. The energy of that light quantum is exactly equal to the difference in the energy levels, in accordance with the Law of Conservation of Energy. The atom loses energy as a consequence and becomes more stable.

(b) Spontaneous emission is the emission of a light photon when an electron in an excited state relaxes by falling to a state of lower energy, irrespective of other radiation which may be present. In stimulated emission an incoming photon of a specific frequency interacts with an excited electron, causing it to drop to a lower energy level. The photon produced has the same phase, frequency, polarisation, and direction of travel as the stimulating photon.

(c) There are generally many more electrons in the ground state than in excited states, making excitation much more probable than stimulated emission.

(d) It is essential to bring about a situation in which there are more electrons in excited states than in the ground state. This is called a population inversion and is brought about by optical pumping. The target material needs to be chosen with care so that it has a metastable state in which excited electrons can exist longer than in a normal excited state. The photon emitted by the first spontaneous emission from the metastable state precipitates a rapid cascade of stimulated emissions.

(e) Laser light is coherent, monochromatic and polarised. Sunlight is incoherent, has a continuous range of wavelengths, and is unpolarised.

5. (a) See text.

(b) The target is embedded in a large mass of copper which conducts the heat away. Oil circulates through the target in thin pipes. In this way heat conducts into the oil and away from the target.

(c) An incoming electron will knock an electron out of the filled electron shells. The vacancy left is immediately filled by an electron from a higher energy shell dropping down to a lower energy shell. The difference in energy corresponds to the energy (and hence the wavelength) of the X-ray photon emitted. Since the energy differences can vary depending on which transition occurs, different discrete lines occur in the spectrum.

(d) (i) $\lambda = hc \div E$
$= (6.63 \times 10^{-34} \times 3 \times 10^8) \div (1 \times 10^5 \div 1.6 \times 10^{-19})$
$= 1.24 \times 10^{-11}$ m

(ii) The wavelengths in the discrete X-ray spectrum would differ.

6. (a) See text.

(b) Cost: CT scans are very expensive.
Radiation: CT scans give a dose corresponding to around 100 conventional X-rays. The doctor must weigh up the benefits of the CT scan against the radiation danger to the patient.
Patient age: Babies are particularly sensitive to X-rays and are generally not given a CT scan.

Exercise 18 – *Unit 2.6*

1. (a) (i) From definition of KE, $v = \sqrt{\dfrac{2 \times KE}{m}}$
$= \sqrt{\dfrac{2 \times 2.2 \times 10^{-18}}{9.11 \times 10^{-31}}} = 2.2 \times 10^6$ m s^{-1}

(ii) $p = mv = 9.11 \times 10^{-31} \times 2.2 \times 10^6 = 2 \times 10^{-24}$ Ns

(iii) $\lambda = \dfrac{h}{p} = (6.63 \times 10^{-34} \div 2 \times 10^{-24}) = 3.3 \times 10^{-10}$ m

(b) $\lambda = \dfrac{h}{p} = (6.63 \times 10^{-34} \div 0.025 \times 400) = 6.63 \times 10^{-35}$ m,

which is much too small to show diffraction.

3. $p = \dfrac{h}{\lambda} = 6.63 \times 10^{-34} \div 3.64 \times 10^{-10} = 1.82 \times 10^{-24}$ Ns

$m = \dfrac{p}{v} = 1.82 \times 10^{-24} \div 2 \times 10^6 = 9.11 \times 10^{-31}$ kg,

which is the mass of an electron. The particle is therefore likely to be an electron.

3. (a) The speed of the two particles is the same, but the electron has the lower mass and therefore it has the smaller momentum. Its smaller momentum means the electron has the longer de Broglie wavelength.

(b) Since the particles have the same speed, the ratio of their de Broglie wavelengths is in inverse proportion to their mass.
$\dfrac{\lambda_e}{\lambda_p} = \dfrac{m_p}{m_e} = \dfrac{1.66 \times 10^{-27}}{9.11 \times 10^{-31}} = 1833$

(c) Since KE = ½ mv^2 and Momentum p = mv, then
$p^2 = 2 \times KE \times m$, so $p = (2 \times KE \times m)^{½}$
Since the KE for both beams is the same:
$\dfrac{\lambda_e}{\lambda_p} = \left(\dfrac{m_p}{m_e}\right)^{½} \dfrac{(1.66 \times 10^{-27})^{½}}{(9.11 \times 10^{-31})^{½}} = (1.66 \times 10^{-27})^{½}$
$= (1833)^{½} = 42.8$

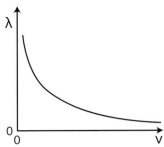

4.

Note the (0,0) origin – so the graph must not touch either axis.

5. KE = ½mv^2 so, $100 \times 1.6 \times 10^{-19} = ½(9.11 \times 10^{-31}) \times v^2$
So: $v = \sqrt{\dfrac{3.2 \times 10^{-17}}{9.11 \times 10^{-31}}} = 5.927 \times 10^6$ m s^{-1}
$\lambda = h \div mv = (6.63 \times 10^{-34}) \div (9.11 \times 10^{-31} \times 5.927 \times 10^6)$
$= 1.2 \times 10^{-10}$ m

Exercise 19 – *Unit 2.7*

1. Maximum frequency $f = \left(\dfrac{v_w}{v_w + v_s}\right) f_o$, so:
$f = \left(\dfrac{330}{330 - 2}\right) \times 250 = 251.52 \approx 252$ Hz

Minimum frequency $f = \left(\dfrac{v_w}{v_w + v_s}\right) f_o$, so:
$f = \left(\dfrac{330}{330 + 2}\right) \times 250 = 248.49 \approx 248$ Hz

2. Use $f = \left(\dfrac{v_w}{v_w + v_s}\right) f_o$:
$88 = \left(\dfrac{330}{330 + v_s}\right) \times 100$
Rearrange to give: $0.88 \times (330 + v_s) = 330$
so, $v_s = 45$ m s^{-1}

3. (a) $z = \dfrac{\Delta\lambda}{\lambda} = \dfrac{(124.2 - 122.2)}{122.2} = 0.01637$

(b) $v = zc = 0.01637 \times 3 \times 10^8 = 4.911 \times 10^6$ m s^{-1}

(c) There is an increase in the observed wavelength (red shift) so the galaxy is moving away from the Earth.

4. $z = \dfrac{\Delta\lambda}{\lambda} = \dfrac{(657.8 - 656.0)}{656.0} = 2.744 \times 10^{-3}$
$v = zc = 2.744 \times 10^{-3} \times 3 \times 10^8 = 8.23 \times 10^5$ m s^{-1}
$d = \dfrac{v}{H_o} = 8.23 \times 10^5 \div 2.4 \times 10^{-18} = 3.43 \times 10^{23}$ m